The Geopolitics o

'Written by some of the most important activists/theorists of the ecological/degrowth/ debt movements … a most powerful and comprehensive analysis of the forces and projects that are threatening the future of our planet. International in its scope, it demonstrates the colonial roots and character of the politics of "extractivism", it unmasks the exploitative character of green capitalism, and highlights the strategies and movements we must support to put an end to the corporate devastation of our communities and the earth. This is a most essential reading for those struggling to create a world where "life is at the centre". Study groups should be organised to spread its knowledge and vision of a different, life affirming future.'
—Silvia Federici, feminist activist, scholar, author of *Caliban and the Witch*

'The big question in critical environmental thinking today is whether the transition to a post-carbon world can take place without a radical overhauling of the system of production globally. The answer this collection of impeccably documented, well-argued studies comes up with is an unqualified no – a post-carbon world needs to be a post-capitalist world.'
—Walden Bello, author of *Deglobalization: Ideas for a New World Economy*

'Drawing on critical feminist, ecological and decolonial perspectives from leading scholars and activists, this book brilliantly surveys this terrain across a broad range of sectors and regions of the world, as well as highlights the resistance grounded in a belief that another world is possible and being built through everyday struggle.'
—Peter Newell, Professor of International Relations, University of Sussex

'As conventional forms of colonialism and neo-colonialism are challenged across the world, capitalist and statist forces are trying hard to sustain their own profits and power (rather than sustain the earth) by seductive but superficial "solutions" like the green economy, carbon markets and technofixes. By exposing these trends for what they are, and by providing genuine, radical alternatives that could lead to a truly equitable, regenerative future, this book is a very valuable contribution to local-to-global struggles for a just world.'
—Ashish Kothari, Kalpavriksh, co-editor of *Pluriverse: A Post-Development Dictionary*

'If "green" ever meant ecological, it certainly doesn't now: the business of the climate catastrophe is an opportunity for transnational profit. This fine collection analyses the contours of power that capitalists have used to extract from the Global South. More than that, through the contributors' range of experiences and struggles, they're able to do what no one author could: flag the many sites and modes of resistance in a way that offers genuine decolonial hope.'

—Raj Patel, Research Professor, University of Texas at Austin

'This volume sheds harsh light on entanglements of "green technologies" with socio-political arrangements that facilitate exploitation toward profit; it also nourishes hope via explorations of liveable horizons illuminated by decolonial, eco-feminist and degrowth perspectives.'

—Susan Paulson, Center for Latin American Studies, University of Florida

'An indispensable read for the actors, policymakers and citizens confronted with the major challenge of our time. The authors show that another approach to the ecological transition is not only possible; it is urgent if we want to live together on a limited planet.'

—Geoffrey Pleyers, President of the International Sociological Association

'Each chapter in this extraordinary book explores and exposes how the North has found a new way of extracting value – I insist: material and moral – from the South. Through the team of authors of this excellent and well-articulated collective work, we witness not only the process of expropriation of goods and resources, but also the usurpation of the vocabulary with which we had named a precious historical goal – ours too – such as the defence of the environment. Double theft: of nature and of the words with which we intended to defend it. The structure of the coloniality of power, untouched by time, re-emerges before our eyes mapped in a new form ... that has rarely been mapped in this way.'

—Rita Segato, feminist decolonial anthropologist

The Geopolitics of Green Colonialism

Global Justice and Eco-social Transitions

Edited by
Miriam Lang, Mary Ann Manahan
and Breno Bringel

First published 2024 by Pluto Press
New Wing, Somerset House, Strand, London WC2R 1LA
and Pluto Press, Inc.
1930 Village Center Circle, 3-834, Las Vegas, NV 89134

www.plutobooks.com

Translations from Spanish to English: Heather Hayes

This publication was partly financed by the Rosa Luxemburg Foundation with funding from the Federal Ministry for Economic Cooperation and Development of the Federal Republic of Germany.

British Library Cataloguing in Publication Data
A catalogue record for this book is available from the British Library

ISBN 978 0 7453 4934 3   Paperback
ISBN 978 0 7453 4936 7   PDF
ISBN 978 0 7453 4935 0   EPUB

This book is printed on paper suitable for recycling and made from fully managed and sustained forest sources. Logging, pulping and manufacturing processes are expected to conform to the environmental standards of the country of origin.

Typeset by Stanford DTP Services, Northampton, England

Simultaneously printed in the United Kingdom and United States of America

Contents

PART III   HORIZONS TOWARD A DIGNIFIED AND LIVEABLE
           FUTURE

Figures

Acknowledgements

Gratitude is owed to all the people and unsung heroes across the globe, fighting in different arenas to sustain life on our planet, and creating a liveable and dignified world for all. Today, more than ever, it is necessary to weave initiatives, experiences, movements and eco-social organisations that unfurl the boundless possibilities and horizons of profound transformations. Our utmost appreciation goes to those building these paths, unfazed by the stormy challenges and violence of this critical juncture.

This book is the result of many collective and global dialogues. There are several organisations, institutions and processes without which this book would not exist: the Global Working Group Beyond Development, the Eco-social and Intercultural Pact of the South, the Institute for Policy Studies (IPS) and all the people and groups that participated in the Global Just Transition process and the South–South dialogues promoted by the IPS and the Eco-social Pact of the South since 2022.

We also extend our gratitude to the Rosa Luxemburg Foundation, which has hosted and financed the Beyond Development Global Working Group between its creation in 2016 and 2022, as well as its Regional Office for West Africa located in Dakar, which hosted the meeting of 2022 on Just Transitions and supported the writing of many chapters of this book.

We also want to thank our families, partners, daughters and sons, and colleagues, who have supported us through compiling, organising, writing and revising this book with their patience, supportive criticism and love.

Some individuals also need to be explicitly recognised. John Feffer has been a vital supporter of the process behind the book and its publication. Claus-Dieter König and Ibrahima Thiam have allowed many authors to meet in Senegal to discuss preliminary ideas for the book. Ken Barlow and the entire Pluto Press team enthusiastically welcomed our proposal, ceding the copyright to translate the manuscript into several languages. Last but not least, we want to give exceptional thanks to Heather Hayes, who soberly and professionally translated different chapters of the book from Spanish to English.

Lucrative Transitions, Green Colonialism and Pathways to Transformative Eco-social Justice

An Introduction

Miriam Lang, Breno Bringel and Mary Ann Manahan

The challenges human societies face are growing more complex at a breathtaking pace. Over recent years, multiple disasters related to global warming, the pandemic and war in Ukraine have further complicated a panorama already tensioned by contending geopolitical powers. Scientists have developed enormous skills in spotlighting and measuring those challenges, in modelling and predicting the future with scenarios of global warming and species extinction. They provide us with information as detailed as never before. However, despite knowing more than ever, we are struggling hard with adequate responses to this situation. According to science, the remaining timeframes to address ecological collapse are shortening, but greenhouse gas emissions are still on the rise after decades of multilateral efforts. Like the pollution and acidification of oceans, loss of species, freshwater and soil fertility, all the indicators of environmental destruction are relentlessly getting worse, and we seem doomed to expand our social metabolism even further on a limited planet.

The very instruments human societies have developed to address these challenges have themselves become part of what has been characterised as a polycrisis, that is, a series of interconnected crises that reinforce each other. The institutions of liberal democracy are facing increasing loss of credibility, while authoritarian forces also dismantle them from within in many parts of the world. In terms of wealth distribution, our world has never been as unequal as today, yet we find no means to put limits to the skyrocketing fortunes and influence of a handful of people and transnational corporations.

All of this has severe effects on the decisions we make regarding our relations with Nature and their outcomes. Environmental politics today are shaped by two characteristics: first, they are not primarily focused on pre-serving complex ecosystems, but on accumulating capital. And second, they are colonial in scope: that means they assume that some regions of the world, some bodies and populations need to be of service to others when it comes to environmental conditions that allow a life in dignity. In the following, we will dive deeper into these two arguments that are at the core of this book.

'DECARBONISATION CONSENSUS' AND PROFITABILITY AT THE CORE OF ENVIRONMENTAL POLITICS

Environmental and climate governance have been no exception to the impo-sition of neoliberal reason onto all dimensions of life after the 1980s. As Laval and Dardot have pointed out, neoliberal reason is not content with removing any limits to businesses in the name of freedom, but it aims to extend the very business logic and its mode of operation far beyond the sphere of the market, into all areas of society and even subjectivities – putting the state at work to become the main lever of this extension.[1] Neoliberal reason reshaped the boundaries of what kind of policy proposals were acceptable, or even thinkable, and it did so quite successfully as it became hegemonic.

This is how profitability came to be at the heart of environmental politics. Shortly after environmental movements and academic reports such as 'The Limits to Growth',[2] borne out of the concern about the obvious conse-quences of the golden decades of Fordism for the environment and human health, had pushed through a relatively effective system of rules and thresh-olds, lawsuits and sanctions against polluters at least in the Global North, it was replaced by a system of market mechanisms that actors would adopt 'voluntarily', simply because they were profitable.[3] Since the Kyoto Protocol in 1997, carbon offsets, clean development mechanisms, etc. have prevailed, a set of market incentives that have often changed in name since then, but not in their underlying logic. The protection of our habitat has become the subject of speculative deals that end up financialising the atmosphere itself. They often only simulate a reduction in greenhouse gas emissions by making assumptions about who would otherwise have cut down a forest, for example, if it had not been included in a carbon offset deal. At the same time, the language of climate change has become so complicated that it has effectively excluded many grassroots actors and made pollution an expert issue. The hegemonic climate discourse today stages a gigantic simulation in which 'sinks' on one side of the planet supposedly absorb certain tonnes of CO_2 emitted on the other side, as if those 'sinks' had not always absorbed carbon anyway, and as if those emitted tonnes of CO_2 would not add to pollution in absolute terms.

In recent years, a new global agreement has emerged, committed to transforming the energy system from one based on fossil fuels to one with reduced carbon emissions, based on 'renewable' energies. Its leitmotif is to fight global warming and the climate crisis by promoting an energy transition driven by the electrification of consumption and digitalisation. However, instead of protecting the planet, it contributes to its destruction, deepens existing inequalities, exacerbates the exploitation of natural resources and perpetuates the model of the commodification of Nature. Breno Bringel and Maristella Svampa define this process as the 'Decarbonisation Consensus'.[4]

On the one hand, it suggests that everything could go on as before if only we replaced fossil fuels with renewable ones. On the other hand, it insists on the centrality of economic growth to the organisation of our economies and societies, growth that would now simply be 'green'. Moreover, the Decarbonisation Consensus limits the horizon of the fight against climate change to what the Brazilian researcher Camila Moreno defines as 'carbon metrics':[5] a limited way of quantifying carbon, based only on CO_2 molecules, which provides a kind of currency for international exchange and creates the illusion that something is being done about environmental degradation. These metrics reduce the deterioration and increasing collapse of the enormously complex web of life on Earth to one single number, easily compatible with the capitalist ratio of accounting: tons of CO_2, as if this figure could provide reliable information about the manifold damages caused to our habitat by the hegemonic mode of living and their interconnecting lines. This hides the underlying problems and explicitly advocates 'green business', 'nature-based solutions', 'climate-smart mining', 'carbon markets' and various forms of speculative investment. Although the seriousness of the climate emergency is recognised in principle, policies are being constructed that are not only inadequate but also have very serious consequences.

In recent years, all major world powers (the EU, US and China) have made commitments to reduce carbon emissions and shift their economies toward low-carbon and decarbonised modes of production, aiming simultaneously at new opportunities of 'green' economic growth. But this Decarbonisation Consensus is marked by green colonialism. It mobilises both neocolonial ecological practices and imaginaries. Under a new twist on the rhetoric of 'sustainability', a new phase of environmental dispossession of the Global South is opening up, affecting the lives of millions of human and non-human sentient beings, further compromising biodiversity and destroying strategic ecosystems.

The war in Ukraine lately has remapped geopolitical tensions and reinforced international dependencies on fossil fuels, as it prioritised short-term concerns for energy security. Transnational oil and gas corporations simultaneously plan to expand their fossil fuel operations while they also explore lucrative new technologies, for example around hydrogen. 'Green capitalism' or 'green extractivism' is the name now given by activists and much of the academic literature to the dynamics of 'accumulation by defossilisation'.[6]

Technological research and innovation are thriving, but they too are deeply inscribed in the paradigms of profitability, infinite progress and economic growth, instead of being oriented by the fundamental need to sustain and reproduce life. Thus, they are mainly directed at even more thorough destruction of our habitat and social fabric, opening paths to the exploitation of hydrocarbons in more and more risky settings, to geoengineering at a planetary scale to get temperatures 'back under control', or to artificial intelligence to replace human learning and understanding of the complex interrelationships that are at the origin of life itself. This is why so many voices from the Global South today denounce hegemonic transitions and their logics, even if their voices are not always heard.

A GLOBAL SOUTH MADE INVISIBLE TO BE APPROPRIATED

As persons whose lives and struggles gravitate in the Global South, but who have multiple bonds, and for some of us, also roots in the Global North, we have been witnesses to many debates and practices that pretend to address ecological collapse in different world regions. Especially in the US and Europe, we are struck by the consistent invisibility of the Global South in these debates, by how naturally it is assumed that all the 'critical raw minerals' and extensions of lands necessary for all the electric cars, the giant solar or eolic installations, the digitalisation of production which are being promised in order to achieve green growth, will come from somewhere. The documents rhetorically focus on 'green alliances' and 'sustainable raw materials'[7] to outsmart the other world powers in the race for geoeconomic primacy, without giving any detail on how extractivism will turn 'sustainable' and North–South relations less asymmetric. Their concerns are centred on the needed quantities.

Meanwhile, in the Ecuadorian tropical forest, deforestation is being pushed by Chinese appetites for the extremely light balsa wood tree, which is used in the building of wind turbines. In South Africa, huge infrastructures for hydrogen plants for export of 'clean' energy become a predicament for communities who ground their living on small fisheries or agriculture. In the Maghreb, pastoralists lose their lands and water to huge solar farms which are built to provide 'green energy' to Europe. In the South American lithium triangle, indigenous communities struggle for the scarce water sources that are increasingly being grabbed by lithium mining in order to equip all the electric cars with lithium batteries. All these recent practices of appropriation and dispossession are labelled as 'green', which provides them with a whole new legitimacy in the face of today's struggles for livelihoods or territories.

Green colonialism unfolds in at least four different dimensions of the relations between the geopolitical Norths and Souths as they are being reshaped and updated in the context of this Decarbonisation Consensus. First, in the claim on unlimited raw materials in the context of 'resource security' policies, which adds an extra 'green' layer to the already existing extractivist pressures. Second, in the imposition of certain formats of conservation in Southern territories in the context of carbon offset schemes, which at the same time allow to further postpone urgent structural changes in polluting production processes located in Northern economies. The third dimension is the use of places in the Global South as dumpsites for the toxic and electronic waste generated using renewable energy sources;[8] and the fourth is projecting the South as new markets to sell new renewable technologies at high prices within the asymmetric architecture of global trade, thus perpetuating unequal exchange.[9]

We are struck by how easily in many debates of the Global North, the geographies where that appropriation will take place are imagined or represented as without people or conflict, as if they were on another planet altogether where nothing should be of concern. We are incensed at how certain landscapes, bodies and whole populations in the Global South are rendered disposable. This is how the mutual constitutiveness of colonialism, patriarchy, racism and capitalism that have existed since the sixteenth century are re-enacted today: geographies destined for accumulation prey on other geographies, destined to be plundered.[10] While today's green colonialism is exactly as materially expropriating as other colonial relations before, it complicates resistances by proclaiming itself environment-friendly and indispensable in order to grant humanity a future, a journey where the racialised populations of the Global South apparently still have no seat.

These practices are continually fed by neocolonial imaginaries. For example, the idea of 'empty space', typical of imperial geopolitics, is often used by governments and corporations. In the past, this idea, which complements the Ratzellian notion of 'living space' (*Lebensraum*), generated ecocide and indigenous ethnocide, and later served to promote policies of 'development' and 'colonisation' of territories. Today it is used to justify territorial expansionism for 'green' energy investments. In this way, large tracts of land in sparsely populated rural areas are seen as 'empty spaces' suitable for the construction of windmills or hydrogen plants.

We are perplexed by how even political forces who consider themselves alternative, dissident or left, that means, in opposition to the hegemony of global capitalist civilisation, also fail to think of eco-social transformation in truly global terms, in a perspective of global social and environmental justice. We are disconcerted by how deeply the habit to externalise the social and environmental costs of an imperial mode of living,[11] sustained both by a normalised day-to-day routine and by historically asymmetric global structures and rules, has become naturalised; and how tightly this frames the corridors of what is politically sayable, let alone feasible. In the discussions about energy transition, efficiency and security, privilege is as strikingly self-evident in the societies of the North as it was during the first years of the COVID-19 pandemic. This self-evidence grounds in the naturality of having grown up in a context where your life and rights are worthy to be protected, and of implicitly being aware that this is not the case for the majority of the world population. The coloniality of being, of power and knowledge, as it was so brightly brought to light by Peruvian sociologist Anibal Quijano and the Latin American working group on coloniality/modernity since the 1990s, is looming everywhere in the debates about energy transition, efficiency and security in the different parts of the North.

This book sets out to make all this visible, and to showcase voices that are usually not heard in these debates. It seeks to amplify the perspectives of networks, movements and alliances in the Global South to help generate a stronger counterweight to the new hegemonic phase of tech-based and corporate-led green colonial capitalism. Intervening in this debate from an intersectional and internationalist perspective, it intertwines strands of debate that otherwise would be split by disciplinary and national constraints.

One of the premises of this book is that there can be no eco-social transformation without global justice. Our planet is a mega-complex ecosystem of which humans are a part. The COVID-19 pandemic has clearly shown us where we end up when we do not consider systemic solutions for all from the outset, but prioritise national or corporate interests. At the same time, we embrace justice in all its dimensions: social, racial, gender, ecological, inter-ethnic or inter-species, as reflected in the diversity of approaches included here, ranging from eco-feminism to ecological economics, and from eco-socialism to pluriverse. All the authors have backgrounds that combine activism with knowledge production in a variety of settings. They do not write about struggles for eco-social transformation, but from within these struggles.

A second central premise is that eco-social transformation needs to urgently reduce human consumption of energy and matter *in absolute terms*, which implies planned, deep changes in our modes of production and provisioning. A planned degrowth, especially in the Global North – accompanied by structural reforms toward a fair distribution of the material means necessary to reproduce life, both within and between countries or regions – is one unavoidable dimension of this transformation. This is why we have invited voices from the degrowth movement to contribute to Part III. Global justice will only be achieved if critical voices from the Global North and South row together, despite their differences, on a shared path.

The authors of this book speak very different languages, not only in terms of their socio-geographical backgrounds, but also in terms of their epistemic and activist trajectories. This diversity is reflected in the style of the chapters, which together form a dialogue between different ways of thinking, knowing and understanding eco-social transformation. We believe that this diversity of approaches is precisely what is needed to overcome the blind faith in technology-based solutions that has brought us to the brink of collapse.

CLIMATE COLONIALITY AS THE LATEST STAGE OF GREEN COLONIALISM

Although the idea of green colonialism has gained traction during the last decade to define the current phase of green extractivism, it was previously used mainly by environmental historians to define a long-term process. As claimed by Richard Grove, one of the leading exponents of this field of study:

> the kind of homogenising capital-intensive transformation of people, trade, economy, and environment with which we are familiar today can be traced back at least as far as the beginnings of European colonial expan-

sion, as the agents of new European capital and urban markets sought to extend their areas of operation and sources of raw materials.[12]

Understood in this way, green colonialism is not a recent phenomenon, but is rather associated with a historical pattern of colonial power and capitalist expansion. Extractivism has been in the DNA of colonialism since 1492. In a brilliant book, Horacio Machado Aráoz shows in detail how Potosí became the starting point of a new era, geological and civilisational, in which modern-colonial mining serves as a trigger for the capitalocene.[13] This pattern has changed over the centuries. While the extractivist logic and colonial violence against bodies, territories and ecosystems have always been maintained, it has become more complex with the emergence of new material conditions and means of justification. As colonialism expanded, a new modern geopolitical imaginary was formed about Nature and the non-Western 'other', justifying land grabs and the subjugation of entire populations. Paradoxically, it was the ecological destruction caused by colonialism that served from the mid-seventeenth century onwards for the emergence of a concern for environmental conservation.

Since then, the colonial powers have complexified their imperial strategy: they continue to destroy Nature and extract as much riches as they can, but at the same time construct conservationist policies and discourses. Vimbai Kwashirai, for instance, has analysed 'green colonialism' in Zimbabwe from the late nineteenth to the late twentieth century and shows in detail both the socio-environmental impacts of British colonialism, and the different types of conflicts, relationships and mediations between colonial officials, corporations, scientists and local actors over timber exploitation and forest conservation.[14] As Ravi Kumar argues, the tension between advocating the conservation and destruction of forests in Africa and Asia is a legacy of British colonialism. In the specific case of southern India, he shows how British 'green colonialism' first destroyed forests – while blaming the natives for doing so – and then created a policy of controlling forest landscapes, arguing that it was important to maintain and extend state control over Nature to control the climate and irrigation systems and thus, improve the country's well-being.[15]

Likewise, technological control and domination of landscapes have been central to the continued reproduction of green colonialism. Daniel Headrick suggests that hydraulic engineering was a major driver of European imperialism.[16] The construction of canals, widening works and dams would have served to maintain imperial power even after formal colonialism through the induced need for technology transfer. But it is not simply a matter of establishing a relationship of material dependence. Worster provides an interesting example, arguing that after the installation of irrigation projects in India and the establishment of various forms of water control, the relationship of the Indian people to water was never the same.[17] Community water systems in different parts of the Global South were thus dismantled and began to be controlled by colonial capitalism and state authorities for their own purposes. Consequently, anthropocentrism implies not only the modern obsession with human control over Nature, but also a form of indifference, disregard and inferiority towards the value of other forms of organisation of social reproduction.

Green colonialism was thus historically forged with capitalism and the commodification of Nature, combining material expansion and subjective control, which is expressed in the 'coloniality of Nature'.[18]

According to Hector Alimonda, one of the driving forces behind Latin American political ecology, for global hegemonic thought and the ruling elites, this coloniality of Nature presents Latin America (and other regions of the Global South) as a subaltern space that can be exploited, destroyed and reconfigured according to the needs of the ruling regimes of accumulation. This affects the biophysical reality (the flora, fauna, human inhabitants, the biodiversity of its ecosystems), the territorial configuration (the socio-cultural dynamics that articulate these ecosystems and landscapes) and the mentalities (coloniality of mind and knowledge).[19]

As green colonialism did not end with the end of formal colonisation, the conceptual differentiation proposed by Quijano between colonialism and coloniality[20] is relevant to differentiate between specific moments and places where imperial domination took place and the colonial matrix of power that persisted after the political independence of the former colonies. Moreover, the framework of coloniality is important to understand how the imperialism of some countries such as the US did not need colonies to exercise its pattern of power and potentiate green colonialism through military threats, the imposition of global markets and other mechanisms of indirect cultural, legal and political rule.

In green colonialism dwells an 'imperial reason'. It is therefore important that future work explores the relationship between green colonialism and ecological imperialism in more detail. Would they be synonyms? Are they rather interdependent but distinct phenomena? A growing literature today, mainly Marxist, has rescued the debate on ecological imperialism, alive in academic debate since the 1980s, emphasising the ecological contradictions of capitalism and the metabolic rift.[21] In a complementary way, other scholars ask rather how ecological imperialism is rooted in everyday practices and supported by institutions. How is this normalised in a way that hides the imperialism it entails? This is what Brand and Wissen call 'the imperial mode of living',[22] which comes very close to what Slater had defined as 'imperiality', that is, 'the perceived right, privilege and sentiment of being imperial or of defending an imperial way of life in which geopolitical invasiveness is legitimized'.[23]

These recent developments are very welcome, as are those that seek to think of degrowth in an anti-colonial political way.[24] They are relevant in terms of North–South relations because they recognise the ecological debt as a central agenda of contemporary struggles, while at the same time claiming the struggle for decolonisation also in the North. However, we need to be careful with a delicate issue: often the anti-imperialist discourse is still widely mobilised against Nature by sectors that call themselves leftist or 'progressive'. Fossil developmentalism is still very present, for example, in certain sectors of the South that claim to defend a just energy transition and at the same time are totally in favour of continuing to exploit oil for the national interest, because otherwise a foreign country would do so. Equally, the idea of the 'right to development' still resonates strongly among actors from the Global South who define themselves as anti-imperialist, even though there is plenty of evidence of ecocide, genocide and epistemic destruction also caused in the name of 'development'.

If in the struggle for decolonisation in Africa, the Ghanaian revolutionary Kwame Nkrumah argued that neocolonialism would be the last stage of imperialism,[25] today we can suggest that climate coloniality,[26] marked by Decarbonisation Consensus, is the latest stage of green colonialism. Saving the climate and decarbonising the economy have become mantras. The historical tension between conservation and destruction is still very much present, albeit with increasingly sophisticated mechanisms of digital and territorial control. In the process, this new form of green colonialism reproduces historical colonial relations and the coloniality of power, but seeks a new social legitimation around the idea of decarbonisation. For some authors it is a new 'carbon colonialism';[27] for others we are facing a 'climate colonialism'.[28] Whatever the nomenclature, there is some consensus within activists and critical thinkers that we are entering a historical tipping point, both in terms of the nature of colonial relations and climate emergency.

HOW TO MOVE BEYOND GREEN COLONIALISM?
JUST TRANSITIONS AND ECO-SOCIAL TRANSFORMATIONS

If we recognise green colonialism and its current facet of green extractivism as an enemy to be fought, it is also essential to discuss how to overcome it. In addition to exposing its false solutions to climate change and critically analysing its impacts, it is equally important to map out and examine what alternatives are available. In recent years, several interesting analyses have been devoted to examining and proposing socio-ecological alternatives from below. The repertoire of interventions is diverse, ranging from environmental and climate justice[29] to ecological grassroots experiences,[30] degrowth,[31] resilience[32] and a wide range of transitional policies.

Often these analyses focus on the local dimension and adaptation to climate change, but rarely on a global justice perspective that considers not only global actions and frames, but also the different worldviews and perspectives behind contemporary struggles. In our case, we prefer to assess localised struggles that are not localist. This means thinking about the resonance between similar struggles in different parts of the world, their scalability, and the possibility of generating articulations and convergences.

Emphasising the importance of the Global South does not imply homogenising the Global North. On the contrary, we need to complexify our analyses in different ways. On the one hand, as we show in this book, green colonialism is not simply something that is imposed from the top-down or from the North to the South. In many cases, what is at stake is also a kind of 'internal green colonialism', which forges the conditions of possibility for the advance of green extractivism based on colonial alliances and relations between domestic and global elites. On the other hand, we also need to value the critical voices inside the Global North and strengthen the links between them and those alternatives from the Global South.

We fully agree with Colombian activist researcher Arturo Escobar when he suggests that in building these bridges between the struggles of the North and the South today, it is necessary to keep in mind several factors[33] such as the importance to resist falling into the trap of thinking that while the North needs to degrow, the South needs 'development' (even if coloured by green tinges). Growth-oriented policies and extractivism are highly destructive of ecosystems and communities and that's why in the last decade, activists and intellectuals in Latin America began to suggest that what matters is not development alternatives, but alternatives to development.[34]

These alternatives necessarily involve the construction of radical, post-extractivist transitions. If the idea of transition – and even just transitions – has been co-opted by capitalism and various institutional actors who use it in a limited and problematic way as a synonym for a market-oriented energy transition, it is important to clarify its meanings and horizons. We believe that eco-social transition needs to be understood as part of a broader process of transformation of culture, economy, politics and society and its relationship with Nature. Furthermore, an eco-social transition cannot be reduced to a promise of future, as it is in the case of hegemonic proposals.

Transitions are already happening in a multitude of experiences in communities and territories, in rural and urban areas, as well as in territorial resistances around the world against green capitalism and its false solutions. We urgently need to map and strengthen these multiple processes of re-existence linked to community energy, agroecological projects, urban gardens, and alternative economies, to name a few. It is in these concrete experiences, which consist of territorialised (eco)utopias, where the most solid alternatives to green colonialism lie.

Amidst this process, we must ask ourselves whether institutional proposals are part of the solution or the problem, as the framework of polit-ical-institutional alternatives, and the diversity of Green New Deals (GND) proposals are also the subject of heated debate. In recent years, a large lit-erature has been devoted to analysing different national cases,[35] although a more systemic,[36] internationalist[37] and global[38] perspective is also present. Despite the diversity of scopes of GND proposals, they share something in common: the need for governments (rather than markets) to lead the energy transition.[39] In fact, in some cases, these institutional-government transitions come close to facilitate and merge with corporate transitions in a dynamic of submission of the public sector to private interests. In a good number of countries, there is a trend towards the formation of major public–private alliances between states and transnational companies, blurring the boundaries between 'corporate transitions' and 'state-oriented transitions', although in rhetorical terms differentiated discourses and spaces may continue to exist. In other cases, however – in a minority, we must recognise – the state claims its autonomy and is more clearly contested, with tensions and forces rejecting its umbilical relationship with the corporate world, pro-posing viable energy transitions that promote economic diversification and decentralisation, and moving closer to the agendas of environmental organ-isations and social movements, as illustrated by the Colombian government led by Gustavo Petro and Francia Márquez. Analysing these mediations and tensions is an important challenge, only partially incorporated in this book, which needs future attention.

In short, alternatives to green colonialism exist not only in intellectual, but also in political and practical terms, although they face extremely complex scenarios. A large part of left-wing thought in the twentieth century, beginning with Marx, understood periods of transition as those in which new social relations emerge within previously existing ones, characterised by the coexistence and struggle between old and new social relations, a struggle in which new forms of relations play a decisive role. Although radical transitions today break with several assumptions of modernity and thus understand eco-social transformation on different bases, it remains essential to analyse this tension between old and new, and between antagonistic forces. As we intend to demonstrate in the following presentation of the sections and chapters of the book, it is our duty to examine the new facets of colonial capitalism, but also to make visible the existing alternatives, considering their contradictions and potentialities.

ABOUT THIS BOOK

The objective of this volume is to unmask the advancement of green colonialism and the underlying destruction and hypocrisy (Part I); analyse the structural and geopolitical entanglements between Global South and Global North that sustain them, and thus must be addressed to build a global justice perspective (Part II); and share a range of perspectives and ideas that constitute milestones on the path to a dignified future. Part III makes both an inventory of practices and knowledges that sustain other, truly sustainable modes of existence which already exist, albeit often invisibilised and unrecognised, and put forward ideas for strategies and policies to be adopted on that path. Throughout this book, authors interrogate the Global North and Global South not as static geographical categories, but as dynamic geopolitical and epistemic constructions situated in both historical and contemporary configurations of power.

In the following, we provide an overview of these sections.

I Hegemonic Transitions and the Geopolitics of Power

The **first part** of the book critically examines the hegemonic project of energy 'transition' in its different dimensions and scales. This part contains contributions that scrutinise converging themes around the 'green fix',

'Decarbonisation Consensus', mutual processes of accumulation and dispossession, the (new) international division of labour and Nature, and other relations between the Global North and Global South. The authors carefully trace the epistemes, goals, actors and interests behind the transition projects in the EU, US and China, and their translations into concrete policies and practices in Latin America and Africa. All the authors share a similar analysis that the 'capitalist techno narrative', a term coined by Svampa, is simply 'old wine in new bottles'. It does not abandon the obsession with economic growth and its unsustainable model of production, distribution and consumption, despite the rhetoric of planetary boundaries.

In dissecting the 'green fix' as a solution to the polycrisis, Kristina Dietz, a German international relations scholar, puts a spotlight on how the energy transition plan towards climate neutrality in Europe heralds a new phase of green extractivism in the Global South. At the heart of this plan is a stern belief in ecological modernisation that advances the restructuring of trade, energy and transportation to fit within the framework of a 'green economy'. She argues that so-called green energy transitions promote a new super cycle of commodities and position resource-rich countries as suppliers of critical resource materials and 'empty spaces' to the North.

However, states in the Global South also have been playing active roles in entrenching their subordinate position in the global supply chain. Argentinian sociologist and philosopher Maristella Svampa narrates how Argentina, Bolivia and Chile, which host the famous 'Lithium Triangle', have aggressively developed national strategies amid a geopolitical race to outrun each other in the lithium market, and in the process, generate a new configuration of global power. Lithium, as Svampa writes, has become a symbolic and material representation and 'skeleton key' for what she calls 'corporate energy transition' to a post-fossil society. Like Dietz, Svampa elaborates the episteme behind this model of transition, which she argues lies in the ideological stance 'that the potential for change is only possible through technological efficiency and consumption without questioning and changing these underlying logics'.

What undergirds this analysis are fundamental political economy questions – *who owns what, who does what, why, and who benefits and loses?* In Hamza Hamouchene's chapter, the Algerian researcher-activist meticulously tells the story of how transitions to renewable energy in North Africa constitute 'energy colonialism', extractive in nature and reproduced in the form of 'green grabbing'. With broad evidence from the region, Hamouchene foregrounds how the colonial tropes about the Sahara are propagated and deployed to implement the EU's green hydrogen strategy under the European Green Deal. His essay also paints a fierce politics of resistance led by sacrificed bodies, induced by the 'Decarbonisation Consensus'.

The contributions to this section all point to the underlying structural causes of the polycrisis. While Hamouchene says that 'capitalism is the culprit', others like Feffer and Lander point to the more specific aspects of capitalism, i.e. the model of production and overconsumption. Several authors point to the continuing colonial relations of power intertwined with the imperial mode of living (Feffer and Lander, as well as Brand and Lang, Dorninger in the second part). All authors illustrate the constitutive relations between dispossession and accumulation at various scales.

In their joint chapter, American policy analyst John Feffer and Edgardo Lander, a Venezuelan sociologist and thinker, ask the question of whether the world's largest polluters can save the planet. They argue that 'the "clean energy" transitions of the US, Europe and China must be assessed not only according to the gap between pledges and targets, stated policies and actual implementation, but also in terms of the overall net harm to the environment and peoples of the South'. Feffer and Lander emphatically point to the externalisation of socio-environmental costs to the Global South, and the refusal of addressing overconsumption in the wealthiest countries as the underlying driver of climate change.

In a complementary way, Ivonne Yanez and Camila Moreno show how the 'Decarbonisation Consensus', as suggested by Bringel and Svampa, is anchored on the premise of false equivalence and obsession over carbon and 'climate neutrality'. Yanez, an Ecuadorian activist, and Moreno, a Brazilian environmental activist-researcher, demystify the idea that one carbon molecule emitted somewhere is equivalent to a carbon molecule sequestered in another place. In an emphatic tone and with evidence from Ecuador, they argue that 'net zero' emissions only provide the licence for the world's polluters to continue polluting.

In the first part of the book, different chapters deal with the relation of hegemonic transition models with time and place. Regarding the latter, the concept of sacrifice zones, initially coined to describe territories annihilated due to nuclear production and testing during the Cold War, has been extended to refer to places and spaces with hazardous levels of pollution and ecological degradation, and where communities have been sacrificed under the guise of economic growth and development,[40] and more recently, energy transitions. The contributions emphasise how the hegemonic transition models render certain lives, bodies, populations and landscapes disposable.

Here, the essays of Hamouchene and Dietz provide graphic descriptions of how indigenous, pastoralist and farming communities are made disposable, and of the responses – protests and resistance struggles that such processes generate. Discontent is a red thread that ties the contributions to this part together.

With respect to temporalities, Feffer and Lander's chapter explicitly refers to the linear timelines and deadlines that are embedded in the stated goals and targets of the various Green (New) Deals. Yanez and Moreno, on the other hand, articulate a different conception of time in relation to transition and pathways out of the polycrisis. They draw from indigenous knowledge where transition means to transform and become, which entails 'walking with the past ahead'. This non-linear way of thinking about time points to a similar narrative put forward by other contributions in the third part of the book: to value other pathways and ways of doing and being as solutions to the climate emergency and polycrisis.

II   Analysing Green Colonialism: Global Interdependencies and Entanglements

Imagining and building eco-social transformations is hampered by structural and geopolitical processes, relations and institutions that entangle the Global North and Global South in an unequal international division of labour and colonial patterns of power. Part II focuses on these often perverse interdependencies and structural factors which complicate the perspective of just and sovereign transitions within places and regions of the Global South.

The main structural obstacles and entanglements addressed in this part are different dimensions of debts, the ongoing dynamics of unequal exchange, the dominant logics, rules and institutions of global trade as well as global environmental governance, and some forms of internal colonialism that also complicate processes of democratic eco-social transformation.

The second part of the book opens with a chapter written by Christian Dorninger, an Austria-based interdisciplinary researcher. He takes a deeper look into the unequal commercial and ecological exchange in trade and economic production and consumption, which he argues is key to understanding the prolonged inequalities and interdependencies between richer and poorer regions of the world. Using footprint indicators, Dorninger shows how the global patterns of ecologically unequal exchange and drain from the Global South through imperial appropriation did not stop with the end of colonial rule. His chapter proceeds to reveal the gargantuan scale of extraction and appropriation that the Global North has extracted and appropriated from the Global South since the 1990s.

In turn, three scholar-activists from Ecuador, Miriam Lang, Alberto Acosta and Esperanza Martinez, draw our attention to the centrality of debt, in its various forms, as 'a powerful means of exploitation, subjugation and enslavement' that gave shape to the international division of labour and Nature, and still structures the relations between the Global Norths and Global Souths, in which the prosperity of some is built on the looting and subordination of others. In their comprehensive take, the authors tackle the different dimensions of debt and their intersections – from sovereign and popular to colonial and ecological debt, stressing the multi-dimensional and devastating impacts on households, the environment, and entire societies. Using intersectional and decolonial feminism, they outline political steps to confront the 'eternal debts of the South'.

The role of the state remains crucial in understanding how debt (re) structures North–South relations. Ulrich Brand, an Austrian scholar, and Miriam Lang, a feminist decolonial scholar, explore the logic of the state as an actor that reproduces capitalist, patriarchal and racist relations, but also as a 'major interlocutor' when it comes to socio-ecological concerns. In their fine-grained analysis, the authors unpack the multiple roles, internal contradictions and ambiguities, and relational and multi-scalar characters of the state, which are key in understanding the complexities that undergird political strategies around dealing with the state in eco-social transformations. In a provocative tone, they invite readers to ponder on the collective challenge of achieving transformative change by focusing not only on changing policies, but also on altering the state apparatus itself, along with its structures, processes and bureaucracies.

In his chapter, Nigerian poet and environmental activist Nnimmo Bassey builds on Brand and Lang's analysis of the state as a 'condensation of societal relations of power' and how colonialism and capitalism have shaped state structures and epistemes. From a Pan-African perspective, Bassey paints a grim picture of green colonialism in the continent – the 'continued collective imagination of African states and their leaders' to depend on fossil fuel revenues that will allow them to adapt to climate-induced disasters. The firm ideological, unwavering belief in foreign direct investments in exchange for natural resources and cheap labour by African states, continues to entrench them in this 'unequal commercial and ecological exchange', in the words of Dorninger.

This global supply chain of critical raw materials and geopolitical scramble are key themes that Indonesian lawyer Rachmi Hertanti examines in her essay. While exploring the role of the 'state' in the North–South entanglements, she centres her analysis on the positions that core-industrialised and peripheral countries hold in the supply chain. She strongly argues that free trade and investment agreements are concrete mechanisms that lock in resource-rich countries in Latin America, Asia and Africa to become constant suppliers of raw materials for the voracious green transition needs of the US and EU.

The multi-scalarity of the entanglements between the Global Norths and Global Souths is reinforced through global environmental multistakeholderism, critically examined by Mary Ann Manahan, a Filipina feminist activist-researcher. She offers a critique of 'multistakeholderism as a form of privatised global governance, marked by corporate capture, democratic deficit and a complicit UN'. Manahan locates the rise of multistakeholderism in global governance spaces within the politico-historical dynamic of the neoliberal counterrevolution of capital that brought significant changes within the UN system, which led to a crisis of multilateralism. Her chapter resonates with Yanez and Moreno's exploration of the epistemes behind the new 'green fix' of 'nature-based solutions', that unfortunately shape the boundaries of solutions to eco-social problems in the world today.

The contributions to this part also offer diverse perspectives on how to transform the unequal interdependencies and North–South relations. Lang, Acosta and Martinez outline possible reparation strategies to confront the multidimensional problems of debt. In turn, Bassey stresses that decolonising transition demands transforming entire energy, economic and political systems through a Pan-Africanism from below. Similarly, Hertanti calls for embarking on a process that develops the praxis of peoples' energy transition led by working-class peoples. Brand and Lang emphasise, however, that democratic structures and processes at local scales are decisive, but insufficient, calling instead for a multi-scalar approach to transforming the state internally and relationally. This also necessitates curbing not only the power of industrialised countries, but also private corporate actors that have become key actors in global governance spaces. As Manahan puts it, reconstituting radical democratic multilateralism requires recentring calls for redistribution of wealth and resources, decision-making powers, and foregrounding the needs and aspirations of the marginalised communities worldwide.

III   Horizons Toward a Dignified and Livable Future

The **third and final part** brings together a wide array of counter-hegemonic approaches and proposals within the Global Norths and Global Souths that build just transitions with the aim of generating socio-ecological transformation. The contributions to this section highlight pluriversal processes and imaginaries that not only unmask green colonialism and challenge official transition projects, but also put into practice the manifold visions, knowledge, relational ontologies and practices that are possible, underway and necessary. By showcasing multi-dimensional horizons rooted in anti-colonial, anti-patriarchal, anti-capitalist and anti-racist struggles, colonised peoples' histories and territories, and different societal Nature relations, the chapters are living testimonies that there are many possible ways to overcome global injustices and green colonialism, toward a world where many worlds fit, as the Zapatistas suggest.

In the opening chapter, Tatiana Roa Avendaño, a Colombian environmental activist, and Pablo Bertinat, an Argentinian electrical engineer and scholar, tackle alternatives to the green energy transitions. Using anti-capitalist and socio-ecological discourses, they contend that 'just and popular energy transitions' recast energy not as a 'sector', but as a collective right and a common central to the fabric of life and the relationships that sustain it. Drawing on indigenous worldviews in Latin America that 'oil is the blood of the earth', such rethinking opens different paths to decommodification, deprivatisation, democratisation and true decarbonisation of the energy system, as well as transforming the production and consumption model that underlays it.

Reverberating across the Atlantic, eco-social transformations are also taken up by Zo Randriamaro, a feminist human rights activist-researcher from Madagascar. Championing a Pan-African decolonial perspective from below, she sketches out the power of African eco-feminisms in shaping radical eco-social transformations in the continent. Her chapter explores the roots, past and present practices, as well as the worldviews of African movements that have embraced eco-feminist politics. The essay offers a politics of hope: that despite the massive colonial exploitation and capitalist expansion, eco-feminist struggles and horizons in the context of African history are transpiring, with women and other marginalised communities on the frontlines.

Degrowth as a political project and proposal is primarily 'developed in and for the core-industrialised countries of the Global North'. This is the counter-hegemonic discourse that Bengi Akbulut unpacks. Akbulut, a Canadian-based feminist political economist from Turkey, eloquently foregrounds degrowth as a proposal that recentres and reorients the economy beyond the notion of biophysical and material downscaling. Using social reproduction as a central organising and mobilising concept for this transformation, she identifies three axes necessary for this: foregrounding a broader conception of what constitutes work; degrowth as/through justice, especially in relation to historical and ongoing North–South relations; and autonomy and democracy as organising principles of a degrowth economy. Akbulut invites readers to dare ask questions about 'what, how and for whom to produce under which conditions', in the hope that it 'will open space for alternative goals and repoliticise the economy by subjecting it to societal deliberation and control'.

Luis Gonzales Reyes takes up Akbulut's challenge in his essay. This Spanish activist, specialising in environment and energy, outlines the necessary profound changes required in terms of breadth, depth and speed to engender eco-social transformations. Using a degrowth approach, he argues for a radical transformation of the economy and the world of work that reduces material and energy consumption, localises and diversifies the economy, integrates production and reproduction into a single unit, and a redistribution of wealth between and within territories anchored on global justice. With evidence from Spain, he demonstrates the urgency of rethinking the concept of work, separating it from 'employment', which has been the basis of the capitalist system, and extending it instead to care and productive community work.

A degrowth economy also requires changing the way we produce food. Bangladeshi NGO leader Farida Akhter shares the experience of the farmer-led movement Nayakrishi Andolon (New Agriculture Movement), which not only resists corporate-driven and chemical-intensive industrial agriculture but advances a praxis of biodiversity-based farming systems involving over 300,000 farming families in the country. Her essay foregrounds the various strategies deployed by this movement, one of which is the reconstitution of community seed networks and knowledge practices led by women farmers.

This interrelated dynamic of resistance and re-existence is foregrounded in the collective chapter of Maria Campo, a Colombian black feminist, and Arturo Escobar, a Colombian researcher-activist and leading scholar, about an ongoing co-construction of radical pluriversal eco-social transformation rooted in a bioregion in southwest Colombia: the Cauca River geographic valley and its territories. Their action research trajectory and journey echoes the other contributions to this part in stressing ecological and relational ontologies that put care and the reconstitution of the fabric of life at the centre. What the authors outline, too, are concrete objectives, strategies and actions that arise from the lived experiences and practices of collectives in the valley, particularly of racialised peoples and bodies that have cared for it, rebuilt and healed the conflict-ridden lands in Colombia.

The contributions so far have either focused on micro- (community-based), macro- (national) or meso- (regional/territory/bioregions) levels. What is obviously missing is a spotlight on the global scale. In the last chapter, Brazilian sociologists and scholar-activists Breno Bringel and Sabrina Fernandes shed light on what they call 'eco-territorial internationalism' as a possible horizon for transformative eco-social justice. The authors locate their proposal within the history of internationalist organising and evolution of internationalisms in recent decades, underscoring the emergence of transnational networks and coalitions, global spaces of convergence and the global justice movement. The chapter provokes readers to rethink the promotion of solidarity among diverse groups affected by the ecological debt and North–South asymmetries, while recognising that there is place for alliances between the Global Norths and Souths.

All the chapters are underpinned by strong anti-capitalist, anti-colonial, anti-racist and anti-patriarchal perspectives, which are key to re-imagining and building counter-hegemonic transition proposals. The contributions to this part foreground several epistemes: the centrality of radically changing the way we produce, distribute and consume – the way our economies and societies are organised; of reforging and recovering relations between society and Nature; and of dismantling internal and international structures and processes of extractive and asymmetrical North–South relations.

However, there are stark differences, and perhaps unexplored, among them, in relation to *scale* (on which level of society to focus); notions of *public* v. *commons*; *sovereignty* and *autonomy* (how to relate to the state? Separate from it or internally restructure it?); the relations between *cities* and the *countryside*, and the role of *tradition* in shaping pathways and horizons to eco-social transformations. It is important for future work to explore these political lines of inquiry more in detail.

Finally, the authors in this collection do not present alternatives as blueprints nor visions to be imposed on other parts of the world, but, as Akbulut writes, rather as 'one among the many other visions of living well and equitably beyond capitalist growth'. This, after all, is the key message we want to emphasise, that a tapestry of alternatives is being woven, in which resistance and re-existence constitute the (re)imagining and building of another world.

NOTES

1. Christian Laval and Pierre Dardot. *The New Way of the World: On Neoliberal Society* (New York: Penguin Random House, 2017).
2. Donella H. Meadows et al. *The Limits to Growth: A Report for the Club of Rome's Project on the Predicament of Mankind* (New York: Universe Books, 1972).
3. Naomi Klein. *This Changes Everything: Capitalism vs. the Climate* (New York: Simon & Schuster, 2014).
4. Breno Bringel and Maristella Svampa. 'Del consenso de los commodities al consenso de la descarbonización'. *Nueva Sociedad* 306 (2023): 51–70.
5. Camila Moreno, Daniel Speich Chassé and Lili Fuhr. 'Carbon metrics, global abstractions and ecological epistemicide'. *Heinrich Böll Stiftung*, Berlin 2015.
6. Ariel Slipak and Melisa Argento. 'Ni oro blanco ni capitalismo verde. Acumulación por desfosilización en el caso del litio ¿argentino?' *CEC* 8(5) (2022): 15–36.
7. European Commission. 'The European Green Deal', 2019, https://eur-lex. europa.eu/resource.html?uri=cellar:b828d165-1c22-11ea-8c1f-01aa75ed71 a1.0002.02/DOC_1&format=PDF (last accessed 20 May 2023).
8. Benjamin Sovacool et al. 'The decarbonisation divide: contextualizing landscapes of low-carbon exploitation and toxicity in Africa'. *Global Environmental Change* 60 (2020).
9. Jason Hickel, Christian Dorninger et al. 'Imperialist appropriation in the world economy: drain from the Global South through unequal exchange, 1990–2015'. *Global Environmental Change* 73 (2022): 102467.
10. Horacio Machado Aráoz. 'Ecología política de los regímenes extractivistas. De reconfiguraciones imperiales y re-existencias decoloniales en nuestra América'. *Bajo el Volcán* 15(23) (2015): 11–51.
11. Ulrich Brand and Markus Wissen. *The Imperial Mode of Living: Everyday Life and the Ecological Crisis of Capitalism* (London: Verso, 2021).
12. Richard Grove. *Green Imperialism: Colonial Expansion, Tropical Island Edens and the Origins of Environmentalism: 1600–1860* (Cambridge: Cambridge University Press, 1995).
13. Horacio Machado Aráoz. *Potosí, el origen: Genealogía de la minería contemporánea* (Quito: Abya Yala, 2018).
14. Vimbai Kwashirai. *Green Colonialism in Zimbabwe: 1890–1980* (New York: Cambria Press, 2009).
15. V. M. Ravi Kumar. 'Green colonialism and forest policies in South India, 1800–1900'. *Global Environment* 3(5) (2010): 101–25.
16. Daniel Headrick. *The Tools of Empire: Technology and European Imperialism in the Nineteenth Century* (Oxford: Oxford University Press, 1981).

17. Donald Worster. *Transformaciones de la Tierra*. (Montevideo: CLAES, 2008).
18. Fernando Coronil. 'Naturaleza del poscolonialismo: del eurocentrismo al globocentrismo', in Edgardo Lander, *La colonialidad del saber: eurocentrismo y ciencias sociales* (Buenos Aires: CLASCO, 2000).
19. Héctor Alimonda. 'La colonialidad de la Naturaleza: una aproximación a la ecología política latinoamericana', in Héctor Alimonda (ed.), *La naturaleza colonizada* (Buenos Aires: CLASCO, 2011), pp. 21–60.
20. Anibal Quijano. 'Coloniality of power, Eurocentrism and Latin America'. *Nepantla: Views from South* 1(3): 533–80.
21. John Bellamy Foster and Brett Clark. 'Ecological imperialism: the curse of capitalism'. *Socialist Register* (2004): 186–201.
22. Ulrich Brand and Markus Wissen. *The Imperial Mode of Living: Everyday Life and the Ecological Crisis of Capitalism* (London: Verso, 2021).
23. David Slater. 'The imperial present and the geopolitics of power'. *Geopolitica(s)* 1(2) (2010): 191–205.
24. Jason Hickel. 'The anti-colonial politics of degrowth'. *Political Geography* 88 (2021): 102404.
25. Kwame Nkrumah. *Neo-Colonialism, the Last Stage of Imperialism* (London: Thomas Nelson & Sons, 1965).
26. Farhana Sultana. 'The unbearable heaviness of climate coloniality'. *Political Geography* 99 (2022): 102638.
27. Kristen Lyons and Peter Westoby. 'Carbon colonialism and the new land grab: plantation forestry in Uganda and its livelihood impacts'. *Journal of Rural Studies* 36 (2014): 13–21.
28. Gurminder Bhambra and Peter Newell. 'More than a metaphor: climate colonialism in perspective'. *Global Social Challenges Journal* (2022): 1–9.
29. Thomas Bond. *Stopping Oil: Climate Justice and Hope* (London: Pluto Press, 2023).
30. Peter Gelderloos. *The Solutions are Already Here: Strategies for Ecological Revolution from Below* (London: Pluto Press, 2022).
31. Mathias Schmeltzer, Aaron Vansintjan and Andrea Vetter. *The Future is Degrowth* (London: Verso, 2022).
32. Jeremy Rifkin. *The Age of Resilience: Reimagining Existence on a Rewilding Earth* (London: St Martin's Press, 2022).
33. Arturo Escobar. 'Degrowth, postdevelopment, and transitions: a preliminary conversation'. *Sustainability Science* 10(3) (2015): 451–62.
34. Miriam Lang and D. Mokrani (eds). *Beyond Development: Alternative Visions from Latin America*. (Berlin: TNI/Rosa Luxemburg Foundation, 2013). (Spanish version, *Más allá del Desarrollo*, published by Abya Yala in 2011.)
35. Kate Aronoff et al. *A Planet to Win: Why We Need a Green New Deal* (London: Verso, 2019).
36. Noam Chomsky and Robert Pollin. *Climate Crisis and the Global Green New Deal: The Political Economy of Saving the Planet* (London: Verso, 2020).
37. Bernd Riexinger et al. *A Left Green New Deal: An Internationalist Blueprint* (New York: Monthly Review Press, 2021).
38. Max Ajl. *A People's Green New Deal* (London: Pluto Press, 2021).

39. Kyla Tienhaara and Johana Robinson. *Routledge Handbook on the Green New Deal* (Abingdon: Taylor & Francis, 2023).

40. Katia Valenzuela-Fuentes, Esteban Alarcón-Barrueto and Robinson Torres-Salinas. 'From resistance to creation: socio-environmental activism in Chile's "Sacrifice Zones"'. *Sustainability* 13(6) (2020): 3481.

PART I

Hegemonic Transitions and the Geopolitics of Power

1

Global Energy Transitions and Green Extractivism[1]

Kristina Dietz

The emerging global energy transition as a path toward decarbonisation, climate change mitigation and energy security points at a new global commodities boom. A growing demand for so-called critical raw materials – such as copper, lithium, nickel and cobalt, which are essential metals for a carbon-neutral transformation of the global economy and national energy systems – is driving up prices.[2] To what extent this development is leading to a new, so-called 'green' extractivism in the regions of the world with an abundance of these raw materials is not only a question of demand and prices. It also depends on the location of resource-rich countries within a new emerging green global division of labour,[3] and on political decisions, institutional regulations, patterns of state–economy interactions and social struggles over the availability and exploitation of these critical resources. In those cases where the state depends to a high degree on revenues from resource extraction and where ecological and social regulation fails to protect ecosystems and political rights, social movements face the challenge of mobilising not only against extractivism, but also against a hegemonic green discourse that makes it ever more difficult to forge international alliances.

The chapter is structured as follows. In the first section policies of green transition in Europe are discussed. A particular emphasis will be put on the strategies to promote the production of green hydrogen as the new panacea for solving the climate and energy crisis. In what follows, the transregional repercussions of these policies are analysed, with a particular regional focus on Latin America. In this section it is argued that green energy transition policies in Europe and elsewhere and the emergence of green extractivism in Latin America are intertwined in complex ways. In the conclusion I sum up the main findings and discuss what these interactions mean for more radical and emancipatory approaches to transformation from below.

GREEN ENERGY TRANSITION

Since the end of 2020, all raw materials included in price indexes have become more expensive, and the costs of critical raw materials, in particular, have been rising rapidly. The reasons for this change are complex: in addition to expectations of economic growth after the COVID-19 pandemic and the impact of the war in Ukraine on the global supply of raw materials, the government and supra-state programs for the energy transition towards 'climate neutrality' announced around the world are also driving profit and price expectations. One of these programs is the EU Commission's European Green Deal.[4] This ecological modernisation project aims to decarbonise Europe's economy by 2050, i.e., to make it carbon-neutral. One of the main ways to achieve this is through massive electrification of the economy and transportation, a task for which metals such as copper or lithium are indispensable. Copper is required to conduct electricity, and lithium is needed to store it in batteries. The International Energy Agency (IEA) predicts that demand for lithium will increase forty-three-fold by 2040 compared to 2020, and this figure will increase twenty-eight-fold in the case of copper.[5] Financial consultancy Bloomberg estimates that copper demand will even increase by more than 50 per cent from 2022 to 2040 and that the global economy will face a copper scarcity from 2035 onwards.[6]

The German government, for instance, is pursuing similar goals with its energy transition. According to the wishes of the current ruling coalition composed of the social democrat SPD, the liberal FDP, and Alliance 90/The Greens, 15 million electric cars will be registered by 2030. With the war in Ukraine, the government has revised its goals to expand renewable energies in the electricity sector. At the same time, it changed the discourse: due to the fact that importing gas, coal and oil from Russia has now become a problem, renewable energies are no longer just a means of climate protection but have become a 'question of national security'[7] and 'energies of freedom'.[8] The EU and Germany's coalition government rely primarily on technology and innovation to deal with the energy and climate crises. While they both intend to obtain some of the raw materials needed for this technology-driven green energy transition from recycling and on-shoring initiatives, i.e., the promotion of particularly lithium mining in Europe itself, the majority will be imported. The countries with the largest deposits of these resources are those whose historical role in the global division of labour has been to be major suppliers of raw materials – mainly countries in Africa and Latin America.

Green energy transitions will not occur within state borders, but 'it is global interactions among the economies of different states that enable such transitions to occur.'[9] Whereas investments, patent innovation, manufacturing and installation capacities in the green energy sector are mainly concentrated in a handful of countries like the US, China, Japan, some Western European countries like Germany, Denmark and Finland and a couple of further Asian countries like South Korea, critical resources are mainly located in African and Latin American countries.[10] The geopolitical economy of green energy transition[11] is shaped by globally structured interconnections that locate resource-rich countries of the Global South, whose accumulation strategy has – historically and in the recent past – been marked by resource extraction and exporting with low processing, in the position of critical resource suppliers and of deployers of so-called 'empty spaces' for the installation of wind and solar parks and green hydrogen plants. One fear is that an ecological modernisation aimed at decarbonisation could reproduce this global division of labour, now dyed green, and spur a new global commodity supercycle,[12] which could result in a new phase of unequal exchange through the expropriation of these materials and natural resources in the Global South. Unlike the previous supercycle in the early 2000s, this time the focus is not only on fossil fuels and precious and industrial metals but also on those lubricants that are supposed to power a green, electrified, high-tech global economy. Besides the critical raw materials already mentioned, this also includes green hydrogen.

GREEN HYDROGEN

Green hydrogen will play an essential role in the green energy transition. The word 'green' refers to the method of production: green hydrogen is produced by using renewable energies, whereas grey hydrogen is produced using fossil fuels. When the resulting carbon dioxide emissions are stored underground, it is referred to as blue hydrogen. Worldwide many countries have adopted green hydrogen strategies or roadmaps. As early as June 2020, the German government adopted a national hydrogen strategy. At that time, the goal was set to produce 14 terawatt hours of green hydrogen in Germany by 2030.[13] However, this is far from sufficient according to the estimated demand. The shortfalls will be compensated by imports from 'developing countries with large amounts of sun and wind' that 'have a high potential for producing renewable energies'.[14] In addition to this strategy, the coalition government is entering into bilateral negotiations to secure Germany's future hydrogen supply. These so-called hydrogen partnerships already exist with Morocco, South Africa, Namibia and Chile. Similar arrangements could soon be made with other Latin American countries such as Brazil, Colombia, Argentina and Mexico.

To guarantee its access to the supply of green hydrogen, the German government and German companies are supporting the construction of green hydrogen plants in the Global South, both with capital investments and public money. Together with other countries, Germany is participating in the Clean Hydrogen Mission, founded in 2021, which aims to boost the development of so-called clean hydrogen worldwide and to reduce the costs of producing and transporting it. Expanding the productive capacities and developing infrastructures for the transport of green hydrogen is also the focus of the Hydrogen Congress for Latin America and the Caribbean (H2LAC) launched in November 2021 by the *Deutsche Gesellschaft für Internationale Zusammenarbeit* (GIZ) in cooperation with the World Bank, the UN Economic Commission for Latin America and the Caribbean (ECLAC), and the EU.[15] There are already several projects for the production of green hydrogen for export to Europe underway. One example is an agreement between the government of the state of Ceará, in the northeast of Brazil, with the German multinational Linde, represented in South America by White Martins, to implement a green hydrogen plant in Ceará. The initiative is a result of the Brazil–Germany Alliance for Green Hydrogen, created in August 2020 by the Brazil–Germany Chambers of Rio de Janeiro and São Paulo to promote partnerships and business opportunities between Brazilian and German companies and institutions.[16] Another example is an agreement between Siemens Energy and other transnational companies with the Colombian oil company Ecopetrol to build a green hydrogen plant in Cartagena. The initiative is supported by both, the German and the Colombian government.[17] These examples show that to achieve a green energy transition, countries – particularly those densely populated in Western Europe, thus depend not only on Research and Development (R&D), on access to patents and manufacturing capacities, but also on access to resources and space, that is, areas for the construction of large-scale wind and solar installations, green hydrogen plants and so on.

The discourse surrounding the multilateral initiatives for the promotion of green hydrogen frames them as a win-win situation. What goes unsaid is that the production of green hydrogen requires the construction of gigantic wind and solar farms, just as electromobility requires the extraction of huge quantities of lithium, copper and other metals – in other terms, lots of embodied land and material (see Dorninger in this book). Countries with abundant natural resources and alleged 'empty spaces' have already seen years of conflicts around issues of use, control, access, conservation and the strain on ecosystems, livelihoods and ways of life. All these will increase as a result of the darker side of green energy transition: green extractivism.

GREEN EXTRACTIVISM

The term 'green extractivism' has been used by activists and scientists to criticise the extraction and capitalist appropriation of raw materials, natural resources (such as solar radiation or wind) and labour, especially in the Global South, for the purpose of the green-technological energy transition. In these cases, 'green' does not stand for an environmentally friendly and socially just use of Nature, but rather for restructuring trade, energy and transportation to fit within the framework of a green economy. In green extractivism, the extraction of raw materials becomes a means to an end, which is why it appears to be compatible with the sustainable development goals and inevitable in order to secure a low-carbon future.[18] The term is meant to be a critique of how the structural preconditions and impacts of the green-technological energy transition further entrench green global divisions of labour and Nature as well as global relations of inequality and exploitation. This concept emphasises the fact that in those regions that are (inevitably) 'sacrificed' for the sake of ecological modernisation, the extraction and appropriation of raw materials for ecological modernisation goes hand in hand with the increased control and influence wielded by transnational corporations, international organisations, Western governments and capital fractions on politics, territories and labour.[19]

Green extractivism is not opposed to so-called neo-extractivism, which emerged as a hegemonic model for trade and development in Latin America at the beginning of the twenty-first century. What characterises the neo-extractivism of the 2000s and 2010s – and what makes it similar to green extractivism – is the extraction and export of (in this case mainly) fossil, metallic and mineral raw materials and agricultural goods with fatal consequences, including ecological destruction and the intensification of social conflicts. Extracting countries become highly dependent on revenues from the commodities sector due to the low amount of value created from this process, which ensures large profits for transnational corporations. This process is propped up by the destruction of alternative resources in rural areas and an increasingly violent enforcement of extractivist projects.[20]

However, the aspect that sets green extractivism apart from neo-extractivism is the discourse used to legitimise it. Because it serves green goals, this method of exploiting Nature is described as climate-friendly, sustainable, progressive and ecologically modern by state, international and private-sector actors, as well as non-governmental environmental protection agencies like those that promote development. For example, the EU touts the H2LAC as an opportunity to link climate protection in Europe with the promotion of an energy transition and sustainable economic growth in Latin America.[21] Thus, the global green energy transition is accompanied by the emergence of a new green development paradigm that links technological solutions to the climate crisis with ecological modernisation and economic development. Besides, governments of the extracting countries are playing a much more active role in green extractivism, as they are also promoting the exploitation of critical commodities and the expansion of renewable energies to open up new sites for accumulation and foster green energy transitions in their own countries.

GREEN EXTRACTIVISM IN LATIN AMERICA

Latin America has vast quantities of copper and lithium and abundant wind and sunlight (see Svampa in this volume). These factors make Latin American countries appealing for the production and export of green hydrogen. Together with the political decisions made in the past decades, the conditions for the emergence of green extractivism in Latin America are already in place. In the 1990s, several governments, under pressure from the International Monetary Fund (IMF) and the World Bank, took measures to liberalise and privatise their agricultural and natural resource sectors, which created the political and institutional framework for selling land and resources on a large scale. Today, further reforms have been implemented to promote private investment in green sectors and encourage the extraction and export of critical raw materials and energy sources. These reforms are supported by international financial organisations such as the World Bank, the Inter-American Development Bank, the United Nations Economic Commission for Latin America and the Caribbean (ECLAC), the EU and national development agencies such as the German Agency for International Cooperation (GIZ).

In recent years, several Latin American countries have passed laws and implemented programs for their own national energy transition, including Mexico in 2015, Argentina in 2015 and 2021, and Colombia and Peru both in

2021. From a global political economic perspective, these various initiatives are to be understood not as single national strategies but as part of a process that emerges from multi-scalar (local, national and global) interactions of mutually reinforcing activities.[22] These activities are being undertaken by local, regional and national governments together with private companies in the Global North and South and international organisations pursuing green energy transition and green capital accumulation, whereby asymmetrical North–South power relations persist. Latin American countries have historically been global producers of critical commodities. In the emerging global green division of labour, many countries of the region will continue to play this role, particularly when it comes to the extraction and export of critical resources for the green transition, e.g., copper and lithium. These two metals' production and export volumes have risen in recent years. An example is copper in Peru. Peru is the second largest copper producer in the world after Chile. Between 2012 and 2019, annual production almost doubled from 1.3 to 2.5 million tons. After production declined in 2020 and 2021 due to the pandemic, the government expected copper production to increase by more than 25 per cent in 2022.[23] Sixty per cent of Peru's export revenues come from the mining sector, which means the country's economy depends on revenues from raw materials. According to current price and demand forecasts, these revenues are expected to continue increasing. China, the US, Germany and Japan, which have the largest capacities for industrial processing worldwide, are the main destinations for the export of Peruvian copper.

The export-oriented production of lithium in Bolivia, Chile and Argentina is also expected to increase given high world market prices and to continue generating income for these countries, albeit in different ways in each case. The Bolivian government under left-wing President Luis Arce is trying – so far unsuccessfully – to facilitate the country's ability to create value by developing its own processing industry. Argentina is different: the country has enormous quantities of hitherto unmined lithium, twice as much as neighbouring Chile could produce. The Argentine government under President Alberto Fernández currently has thirteen new projects scheduled for implementation with transnational companies, in an effort to make a significant contribution to future global security of supply, attract new investment to the country, and expand Argentina's export portfolio. Although the realisation of these projects is uncertain and will depend on to what extent social movements will successfully mobilise against them, only with the expectation that these projects might be successfully implemented, financial analysts already classify Argentina as a new global heavyweight in the lithium market.

In Chile, lithium has mainly been exploited in the Atacama Desert in the northern part of the country. However, the Atacama Desert is not only rich in lithium. It is also one of the regions in the world with the highest amounts of direct sunlight. In the country's south, Patagonia has enormous potential for wind energy. In both regions, large-scale solar plants and wind farms will be used to generate electricity. There are also plans to expand the production of green hydrogen there. Starting in 2030, Chile aims to become the world's largest exporter of sought-after green energy carriers. The former Colombian government of the right-wing conservative president Iván Duque – whom left-wing Gustavo Petro replaced in August 2022 – formulated similarly ambitious plans. For example, offshore wind farms will be built in the Caribbean, and wind and solar farms in the north of the country, in the province of La Guajira. Like Chile, Colombia also wants to expand its export-oriented energy portfolio and export green energy in the form of hydrogen in addition to coal, oil and gas. During a trip to Europe in spring 2022, the former Colombian energy minister Diego Mesa called for support in achieving these ambitious goals, which would require investments of three to five billion US dollars. Preliminary agreements have been signed with the Netherlands, and pilot projects for offshore wind plants with the Danish capital are being prepared.[24]

STRUGGLES AGAINST GREEN EXTRACTIVISM

Protests against the extraction of critical raw materials have been happening in Latin America for years. This trend will continue, even if this extraction is justified by ecological concerns. People's basic needs and natural resources are being sacrificed to meet the global demand for copper and lithium. This is especially true for those sectors of the population that are not counted for creating value within the framework of green extractivism: peasant and indigenous communities. National and local environmental movements, youth organisations, small farmers, indigenous and Afro-Latin American groups are therefore protesting together with international partners against displacement, forced resettlement and the destruction of livelihoods, as well as against air and water pollution and the changes of existing rights concerning the use and access to goods essential for people's well-being. In addition, they demand democratic participation, decentralisation, a fair distribution of profits, adequate compensation and access to jobs (see Roa and Bertinat's contribution to this volume).

These protests are, therefore, about not only the pros and cons of resource extraction but fundamental questions of societal and global power relations. Across the region, movements struggle against social inequalities, poverty, the disregard of rights, the exploitation of labour, the authoritarian imposition of economic projects and the outsourcing of ecological and social costs to those areas of the Global South that resource companies and governments often regard as 'empty spaces' and 'underdeveloped'. The example of copper mining in Peru illustrates these protests' social and political dimensions. Copper in Peru is mainly extracted by transnational corporations in large-scale industrial mines. This type of production is capital-intensive but not labour-intensive. The rural, mostly indigenous population in these mining regions rarely finds work in the mines. Instead, the displacement, lack of compensation and environmental destruction caused by copper mining rob them of their livelihoods and infringes upon their basic needs. This is why they fight back. The Las Bambas copper mine, in particular – operated by the Chinese company Minerals and Metals Group (MMG) – is a site of contestation and was occupied by over 100 members of an indigenous community at the end of April 2022. MMG had to halt its operations at the site. The people who previously lived there lost their land and villages when this mine was constructed. They now demand adequate compensation for the resettlement, investments in social and productive infrastructure that will secure them new sources of income and livelihoods, and an end to environmental destruction. At the same time, the protesters stress that mining companies are cheating local communities.[25]

Resistance is also growing against lithium mining in the Andean salt lakes of Chile, Argentina and Bolivia. The protest movements in these regions also demand the protection of indigenous peoples' rights and sensitive ecosystems. There is also opposition to the expansion of energy production from wind and solar sources to manufacture green hydrogen. In northern Colombia, for instance, indigenous groups are protesting against the construction of mega wind farms in their territories. These regions are already suffering the environmental, physiological and social consequences of export-oriented coal mining.[26]

In the context of the global green energy transition, these areas are subject to a double strain since these are the regions in which fossil fuel extraction often overlaps with green mining projects and large-scale plans to produce renewable energy – energy that, in turn, is urgently needed to produce green hydrogen. With their livelihoods destroyed, the inhabitants of these mining regions are forced to pay the highest price for securing energy supply and solving the global climate crisis, to which they have contributed the least compared to almost anybody else.

CONCLUSION

How the global green energy transition is being shaped and what role Latin American countries and other places in the Global South will play in the emerging green global division of labour and Nature is currently being determined. The rules and procedures of this global transition are formulated primarily by representatives of private capital, international financial institutions, development agencies and banks, the ECLAC, national and international energy agencies, and Global North governments – with the participation of their Latin American counterparts. This is not surprising since the former owns the capital needed to build 'green' production infrastructure and exploit raw materials.

For the moment, it is unclear to what extent Chile, for instance, will succeed in preventing the extreme ecological devastation caused by lithium mining, strengthening the rights of the country's indigenous population, and using strict environmental legislation to make the most of the renewable energy potential, especially at the national level. The same applies to Colombia, where Gustavo Petro made it clear during his election campaign that, as president, he intends to overcome the exploitative extractivist logic of the previous commodities boom. In countries whose governments cannot or will not enforce ecological and social regulations, social movements not only face the expansion of green extractivism. They also must challenge a hegemonic green discourse that makes forging international alliances even more difficult.

The current discourse around a green transition differs considerably from the one used by the proponents of neo-extractivism to legitimise their actions. It is no longer just about development, but about green modernisation, green progress, sustainability and solving the climate crisis, and who could object to all of this? Nevertheless, there are several ways to politicise the exploitative structures advanced through the green energy transition. The reduction of carbon dioxide emissions by means of technology-driven solutions from Europe will lead to more socio-ecological devastation in the countries of the Global South. This is why access to land remains the subject of political struggles and counter-movements. The post-growth movement, eco-socialist approaches and feminist indigenous struggles all provide useful discursive frameworks and practical examples for how a different socio-ecological energy transition could be implemented 'from below'. The task of emancipatory left-wing energy transition politics, as Bringel and Fernandes suggest in this book, is to connect these struggles transnationally.

At the same time, it is crucial to put the green energy transition on a radical reformist path and to – insofar as this is still possible – avert the negative impacts that have already resulted from this process. Lithium will continue to be mined, as will copper, and the production of green hydrogen will continue increasing. However, the decisive question is under which political conditions this will happen. Therefore, in parallel to the development of alternative approaches to the energy transition, an emancipatory leftist intervention should also strengthen those forces in Latin America and the rest of the world attempting to limit the impending green extractivism with strict environmental, democratic and socio-political legislation.

NOTES

1. This text is an expansion and updated version of a text by the author first published in German under the title: 'Energiewende und grüne Ausbeutung. Die Energiewende in Europa kündigt einen grünen Extraktivismus in Lateinamerika an'. Parts of the text were translated by Hunter Bolin for Gegensatz Translation Collective and edited by Mariana Fernández for Rosa Luxemburg Stiftung New York Office. The author is extremely grateful to them for this.

2. 'Critical raw materials resilience. Charting a path towards greater security and sustainability', European Commission, 2020, eur-lex.europa.eu/ (accessed 12 December 2022).

3. Eric Lachapelle, Robert MacNeil and Matthew Paterson. 'The political economy of decarbonisation: from green energy "race" to green "division of labour"'. *New Political Economy* 22(3) (2017): 311–27, https://doi.org/10.1080/13563467.2017 .1240669.

4. 'The European Green Deal', European Commission, 2019, https://eur-lex. europa.eu/resource.html?uri=cellar:b828d165-1c22-11ea-8c1f-01aa75ed71a1. 0002.02/DOC_1&format=PDF (last accessed 13 December 2022).

5. 'The role of critical minerals in clean energy transition', IEA World Energy Outlook Special Report, www.iea.org/reports/the-role-of-critical-minerals-in-clean-energy-transitions (last accessed 16 December 2022).

6. James Attwood. 'A great copper squeeze is coming for the global economy'. *Bloomberg*, 22 September 2022, www.bloomberg.com/news/articles/2022-09-21/copper-prices-fall-despite-signs-of-looming-crucial-metal-shortage? leadSource=uverify%20wall.

7. 'Habeck sieht Chance für Comeback der Solarindustrie'. *Der Spiegel*, 30 July 2022, www.spiegel.de/wirtschaft/robert-habeck-sieht-chancen-fuer-comeback-der-solarindustrie-a-21b76af1-d813-4241-84b0-ffc359f1b86a (last accessed 12 December 2022).

8. RedaktionsNetzwerk Deutschland: Lindner. *Erneuerbare Energien sind «Freiheitsenergien».* Zustimmung von Lauterbach, 27.2.2022, www.rnd.de/ (last accessed 12 December 2022).

9. Lachapelle et al., 'The political economy of decarbonisation', 312.

10. Ibid.

11. Gavin Bridge and Erika Faigen. 'Toward a lithium-ion battery production network: thinking beyond mineral supply chains'. *Energy Research & Social Science* 89 (2022): 102659, https://doi.org/10.1016/j.erss.2022.102659.

12. The term supercycle refers to an extended period of time in which commodity demand drives prices far above their long-term trend.

13. '2020 Federal Report on Energy Research: Research funding for the energy transition', Federal Ministry for Economic Affairs and Energy BMWi, Berlin, June 2020, www.bmwk.de/Redaktion/EN/Publikationen/Energie/ federal-government-report-on-energy-research-2020.pdf?__blob=publication File&v=5.

14. 'Annual Economic Report: Towards an ecological social market economy', Federal Ministry for Economic Affairs and Climate Action BMWK, Berlin, January 2022, www.bmwk.de/Redaktion/EN/Publikationen/Wirtschaft/annu- al-economic-report-2022.pdf?__blob=publicationFile&v=2.

15. See the platform's website, https://h2lac.org/ (last accessed 12 December 2022).

16. See www.h2verdebrasil.com.br/sobre-nos/ (last accessed 16 December 2022).

17. See for further information: www.ecopetrol.com.co/wps/portal/Home/es/ noticias/detalle/alianza-internacional-hidrogeno and www.cancilleria.gov.co/ newsroom/news/dialogo-binacional-politica-reindustrializacion-colombia- basada-energias-renovables (last accessed 16 December 2022).

18. Daniel Macmillen Voskoboynik and Diego Andreucci. 'Greening extractivism. Environmental discourses and resource governance in the "Lithium Triangle"'. *Environment and Planning E. Nature and Space* 2(5) (2022): 787–809, https:// doi.org/10.1177/25148486211006345.

19. Donald V. Kingsbury. '"Green" extractivism and the limits of energy transitions. lithium, sacrifice, and maldevelopment in the Americas'. *Georgetown Journal of International Affairs* (2021), gjia.georgetown.edu/.

20. Maristella Svampa. *Neo-extractivism in Latin America. Socio-environmental Conflicts, the Territorial Turn, and New Political Narratives* (Cambridge: Cambridge University Press, 2019).

21. 'New platform will seek the development of green hydrogen in Latin America and the Caribbean', *PV Magazine International*, last modified 2 December 2021, www.pv-magazine.com/.

22. Lachapelle et al. 'The political economy of decarbonisation'.

23. 'Perú espera elevar producción de cobre en 27% y de oro en 12% en el 2022', Guía minera de Chile, 2022, www.guiaminera.cl/peru-espera-elevar-produccion-de- cobre-en-27-y-de-oro-en-12-en-el-2022-ministro/ (last accessed 12 December 2022).

24. 'Colombia busca hasta 5.000 millones de dólares para hidrógeno verde', Portafo- lio, last modified 13 November 2021, www.portafolio.co/.

25. Carlos Peña. 'Las Bambas. Comunidad campesina de Pumamarca protestó contra empresa minera.' *El Comercio*, 12 February 2021, https://elcomercio.pe/.
26. Eliana Mejía. 'Wayús completan 10 días de protestas en el parque eólico Guajira 1'. *El Tiempo*, 16 January 2022, www.eltiempo.com/.

2

Corporate Energy Transition
The South American Lithium
Triangle as a Test Case[1]

Maristella Svampa

INTRODUCTION

Energy transition is the name we give to the journey from conceiving energy as a commodity and fossil-based, something non-renewable, causing serious impacts on the environment, held privately and concentrated in the hands of a few, to a projection of a common asset that is renewable and sustainable in the full sense of the word, both common and decentralised. This means decarbonising the energy concept and transforming the productive model and, more generally, the system of social relations and connection to Nature. Doing this means abandoning industry-based conceptions and developing a more holistic vision; it also means that the energy transition must be conceived within an entire socio-ecological transition. An energy transition that is not part of a comprehensive vision, one that does not deal with the radical inequality in the distribution of energy resources or lead to decommodification and strengthen the resilience capacities of civil society, will only pay for a partial reform without modifying the structural causes of the socio-ecological collapse that we are experiencing.

However, it is not easy to find this type of systemic transition in the global context. Experiences connected to an energy transition are associated with proposals revolving around changing energy sources, developing renewable energy to replace fossil and nuclear sources, and transforming the energy matrix, but not the system in and of itself. In the energy transition, dominant actors see a potential for wealth accumulation and hegemonic geopolitical positioning, with weak sustainability mechanisms and a corporate outlook. This all could be called the 'corporate environmentalism universe'[2] or the 'capitalist-technocratic narrative'.[3]

As Pablo Bertinat and Melisa Argento maintain, beyond the business sphere, the corporate energy transition may have a wide range of supporters, including multinational companies, states (in their multiple scales), institutions and organisations that support this perspective as the fastest way to respond to the urgency of the crisis, based on the premise of technological efficiency. What matters is sustaining market niches and guaranteeing the supply/demand relationship that exists due to growing consumption, without proposing that the very logic of that consumption be altered.

The objective – which is presented as an end in itself – is to emit fewer greenhouse gases and generate geopolitical support in the face of growing public concern about climate change. This is coupled with a growing accumulation of wealth and power through the new areas of extraction, aiming to maintain existing relations of inequality. At best, it guarantees unlimited growth, but only for a few. This conception is hegemonic, capitalist and colonial to the extent that it promotes false solutions linked to controversial alternatives such as nuclear energy, gas as a 'bridge' or transition fuel, 'extreme' energies, agrofuels, etc.

In the corporate energy transition, most of the elements (artefacts, projects, regulations, research and development, etc.) are controlled by, or work in favour of, transnational corporations or world powers, making systems and everyday life more complex under the excuse of efficiency, and thus limiting the possibility of democratising the use of energy and technology. In this framework, the ownership and control of access to energy sources, materials and necessary technologies plays a central role. The concentration of the energy system is an inherent characteristic thereof. Large companies, not only private, but in many cases public, hold hegemonic power.

The main actors of the corporate energy transition promote the development of renewable energy sources from a utilitarian conception and an industrial format, envisioning that they could be an alternative to the planetary limits of resources within the framework of an intensive neo-extractivist model; in short, dominated by a fossil logic.[4] They imagine that non-fossil energy sources could sustain the current trajectory of unlimited growth. In some instances, a technocratic perspective of the issue associated with energy efficiency also gains prominence. The potential for change is perceived only in technological efficiency and, therefore, in consumption, without suggesting that the very logic of that consumption be altered.

According to a recent study, the media see the energy transition only from an economic and business perspective in Latin America. This bias is clear after examining 1,200 articles from the mainstream media, which Climate

Tracker compiled.[5] The study also highlighted the limited presence of journalists specialised in regional coverage of the energy transition. In addition, a business-centric approach has prevailed (in some countries, overwhelmingly). The primary source of information is national governments, together with corporate representatives. Other noteworthy findings include an absence of scientific explanations, community leaders, and any focus on ecology and poverty.

With this background, the corporate energy transition is based on the questionable idea of 'sustainable development' and the 'green economy', one that revolves around continuing on a path of unlimited growth, exchanging fossil fuels for renewable and high-tech resources, without abandoning models of capitalist consumption or questioning the distribution or access to energy by the people or citizen participation in decision-making processes.

CORPORATE TRANSITION AND LITHIUM

Lithium is considered the skeleton key to the energy transition into a post-fossil fuel society. Lithium is a mineral with various uses, modalities and destinations. One of its primary uses is batteries for personal computers, mobile phones, MP3 players and related products. As a final product, lithium-ion batteries are used for energy storage and to make electric cars. Lithium is also used, among others, to obtain lubricating greases, glass, aluminium, polymers and in the pharmaceutical industry.

Currently, 39 per cent of the demand is for battery production, 30 per cent involves ceramics and glass, 8 per cent for grease and lubricants, 5 per cent for metallurgy, 5 per cent for polymers, 3 per cent for air treatments and the remaining 10 per cent in other heterogeneous uses. Forecasts reflect that by 2026, 70 per cent of lithium consumption will be used for batteries, 15 per cent for ceramics and glass, and the remaining 15 per cent for other uses.[6] Globally, the horizon reflects an accelerated expansion of the market for lithium-based electric accumulators, one that exceeds the market for energy cells and batteries, and suggests that accumulators, which are able to store even more energy, will change the habits of individual consumption and may even come to be required by entire cities to facilitate their energy administration and distribution.

At a geopolitical level, the importance of the lithium market illustrates the new configuration of world power. The need to escape the fossil fuel paradigm and the severity of the climate crisis has generated manifest competition when it comes to obtaining lithium and within the value chain. Very few countries control this chain and can actually manage everything from extracting the ore to manufacturing batteries.

At the beginning of this century, Japan was the market leader in energy cells and batteries, followed by the United States and several European countries, and Japanese firms like Toyota have long sought to be present in the value chain. At that time, China occupied a minimal position (it only represented 1.46 per cent of the total) in this market. However, in an increasingly problematic and complex scenario, while some world leaders (USA, Australia, Brazil, etc.) have focused on denialism, China, until recently reluctant to sign the Kyoto protocol, has modified its policy and appears increasingly on board with the energy transition. This proves that it leads the global electric car market, particularly with BYD, based in Shenzhen, which is even more significant than Tesla in California. It is no coincidence that, in 2017, China became the world's leading lithium battery exporter, on par with the United States and Singapore, followed by Hong Kong. Japan's market share fell drastically, and European countries were providing practically half of the values they had contributed before.[7]

These debates about the role of lithium in the transition, which reflect a repositioning of global powers, are beginning to have an impact in South America, home to what has been known as the 'Lithium Triangle', located between the Atacama Salt Flats in northern Chile, the Uyuni salt flat in Bolivia, and the Salinas Grandes, Olaroz Cauchari and Hombre Muerto salt flats in Argentina. This area concentrates more than half of the planet's proven lithium reserves.[8]

Lithium is an alkaline metal that oxidises rapidly with water or air and has differential properties in heat and electricity conduction. It is present in different types of mineral deposits, as well as in natural brines. The first step in this value chain is to extract lithium carbonate, whether from lithium deposits in brine or mineral deposits (such as Spodumene). While not a rare or scarce mineral, nor unevenly distributed, the most profitable way is to extract it from the Andean salt flats. This means that the global and national pressure on the Atacama region of the salt flats is increasing.

Lithium mining is different from metal mega-mining, since it does not involve removing tons of earth or blowing up mountains; rather, its main problem is that it is, fundamentally, water mining. Its extraction in brine requires the consumption of unsustainable amounts of water in arid regions, which puts the fragile ecosystem of the desert at risk, including its wildlife and the livelihoods of the people who live there, especially indigenous communities. This is what is currently happening in the Atacama region of Chile and Argentina. Due to the consumption of water, the extraction of lithium threatens to break the fragile water balance, tending to dry up aquifers and water reserves in areas already characterised by aridity and water stress. It also competes for water needed by local indigenous community farmers for their crops and for their animals' grazing, while posing a threat to biodiversity. An investigation carried out for Chile by Ingrid Garcés from the University of Antofagasta, found that for every ton of lithium produced, two million litres of fresh water are used, and 'daily, more than 226 million litres of water and brine are extracted from the Atacama salt flat'.[9] This is added to the impacts of nearby metal mining projects, which also extract large amounts of water (Zaldivar and Minera La Escondida). The impact of lithium mining in the Chilean Atacama region led it to become one of the issues of the International Rights of Nature Tribunal, which met in Chile in December 2019.[10]

Lithium mining, in its forms of extraction, production and private appropriation, reproduces the logics behind mega-mining, and in general, the entire aggregate of extractive activities. This all leads to a violation of rights, transferring costs to Nature, territories and populations, which are the parties actually affected by their unsustainability. We are thus facing a model of 'accumulation by defossilisation', as argued by Argento, Puente and Slipak, associated with transnational corporations, in a reproduction of domination over Nature and populations.

In this context, it is hypocritical to appeal to the idea of a 'post-fossil society' or 'energy transition' to demand a specific population's acceptance or turn their territories into zones of sacrifice. Today's consolidated model of transnational mining only serves to justify looting, fuelling the energy paradigm shift in the countries of the Global North, based, once again, on the dispossession of local communities and the destruction of Nature. Among grassroots communities and environmental activists, this is called a 'false solution' since it would serve merely to guarantee what comes to be nothing but a corporate energy transition, which would also benefit the central, and also the richest, countries at the expense of the territories and populations of the South.

ARGENTINA, BOLIVIA AND CHILE: THREE COUNTRIES, THREE DIFFERENT STRATEGIES

According to the United States Geological Survey, 58 per cent of the world's lithium resources and 53 per cent of lithium reserves are concentrated in

Argentina, Bolivia and Chile, in the high Andean salt flats, an area known as the Lithium Triangle, which is home to many indigenous communities. More specifically, 'currently, five ore operations in Australia and four lithium brine operations in Argentina and Chile (2 in each country) account for the majority of the world's lithium production'.[11]

Increased demand for lithium triggered an El Dorado fever, generating different strategies in the three countries involved. Chile is seeking to consolidate itself as the world's largest exporter of lithium without abandoning the privatisation model, meaning that it has developed only the lithium carbonate production phase without including any added value. To this end, Chile has a highly commercialised regulatory framework (which includes the privatisation of water), something that translates into unrestricted state support for mining companies (the two most prominent are SQM and Albemarle) in their need to consume more and more water to produce more tons of lithium for export. During the Michelle Bachelet administration (2014–2018), a National Lithium Commission was created, which recommended looking into sustainability, with the participation of communities and the creation of a state company. None of this materialised, though companies began to pay royalties, which made it possible for the national government to collect revenue. One of the most innovative landmarks has been agreeing with the Council of Atacamenian Peoples, through which Albemarle has agreed to pay the equivalent of 3.5 per cent of sales. According to the Latin American Observatory of Environmental Conflicts, 'shared value adds complexity to trends in Corporate Social Responsibility, moving from a welfare mindset to one of a "participating partner", which seeks to involve communities not only as beneficiaries of the company's profits but also as parties responsible for the impacts that such profits may have while disposing of such people's rights to their territories'.[12]

Bolivia, another quintessential mining country, understood that lithium was not just another commodity but a strategic asset. In response, it looked to generate a more long-term vision based on state control and the industrialisation of lithium. As a result, since 2008, the state has been exploring partnerships with different transnational companies to advance in the subsequent phases (II and III) and produce lithium batteries in the future, ensuring technology transfer and the use of patents. This led the country to make nearly no progress in phase I, lithium extraction, despite having the world's largest reserves in the Uyuni salt flats. Similarly, the Evo Morales administration sought to reach agreements with local communities, especially with the influential mining sectors of Potosí. Along these lines, in 2018,

Yacimientos Litíferos Bolivianos (YLB) was created, 'a state company that includes prospecting, exploration, production, beneficiation or concentration, installation, implementation, start-up, operation, and administration of evaporite resources, inorganic chemical complexes, industrialisation, and commercialisation'.[13]

The overthrow of Evo Morales in November 2019 cut short the possibility of carrying out this ambitious project, which had already triggered a severe conflict between the mining organisations of Potosí and the up-and-coming leaders of the Civic Committee, who had forced the government to back down from the agreement entered into between YLB and a German company, claiming that it left meagre royalties to the area. There is no guarantee that these projects, which aim to achieve the industrialisation of lithium based on the transfer of technology by transnational actors, will be resumed in the future.

In Argentina, lithium extraction has grown exponentially in recent years, with lithium carbonate exports increasing from 8 per cent in 2012 to 16 per cent in 2016, representing growth of 100 per cent in five years. It is currently the third largest exporter in the world, behind Chile and Australia. This trend of acceleration points to increased water consumption. According to forecasts made by Gustavo Romeo, an annual increase of 50 billion litres of water is on the horizon, something equivalent to the total annual consumption of a city of 350,000 inhabitants.[14]

Mauricio Macri's administration (2015–2019) placed the issue of renewable energy on the political agenda. However, Macri did so in a framework of extreme commodification and accentuation of economic and technological dependence. This did not mean that his government had developed any sort of state policy on lithium. Beyond contributing to the El Dorado fever for the mineral known as 'white gold', the involvement of the national government consisted of providing more advantageous conditions to attract mining corporations to Argentina over its neighbours, nations that also have lithium reserves. This accentuated the economic dynamics specific to the relationship between the ore and the state, generating a concession granting system that aggravated disputes over water in arid zones and contempt for local communities. Lithium's appearance as a new business opportunity for the economic and political elite (through the creation of mining service companies or junior companies, to obtain mining claims that are then sold to large industry conglomerates) deepened the unholy alliances between the private and the public.

The fact is that in Argentina, lithium does not have its regulatory framework, as opposed to Chile and Bolivia. Its production continues to be based on the neoliberal regulations of metal mega-mining, which date back to 1990 and includes extensive exemptions, meagre royalties (3 per cent), low withholdings, and the self-exclusion of the provinces to extract the mineral.

Early on, the exploration and production of lithium generated eco-territorial conflicts with indigenous communities. In November 2010, a Board of 33 Native Communities for the Defence and Management of the Territory (Salta and Jujuy) filed an amparo action before the Supreme Court of Justice of the Nation demanding prior consultation on the concession in Salinas Grandes in accordance with the National Constitution (art. 75.17), the General Environmental Law of 2002, ILO Convention 169, and the United Nations Declaration on the Rights of Indigenous Peoples.

Finally, in January 2013, the Supreme Court of Justice of the Nation rejected the amparo, due to a question of jurisdiction. Faced with this, the members of the organisations decided to go to the Inter-American Commission on Human Rights and worked together with several NGOs and recognised foundations for environmental protection (Farn, Fundación Boell), and human rights (Endepa), to prepare the First Indigenous Consultation Protocol of Argentina (2015).

However, the indiscriminate expansion of the mining frontier, the dispute over water, and the absence of prior, free and informed consultation with the communities led to a complicated scenario. For example, in Jujuy, the extractive progress that can be seen in the Olaroz and Cauchari salt flats contrasts with the situation in Salinas Grandes, where their rejection of lithium mining predominates (in 2019, two multinational companies had been awarded contracts). In all cases, the result has been the consolidation of an extractivist scheme similar to that of metal mining, fully transnationalised, highlighted by disputes over water, with a provincialisation of conflicts and the dispossession of populations made up of, above all, native peoples.

Both in Argentina and the Atacama region of Chile, water consumption related to lithium extraction threatens to break the fragile water balance, drying up aquifers and water reserves in areas already characterised by aridity and water stress. It also competes for water needed by local indigenous farming communities and threatens local biodiversity. In many cases, this activity advances without a social licence, meaning without the agreement of the communities. On the other hand, in Bolivia, the lithium industrialisation strategy encountered other obstacles and limits (in terms of extraction and consolidation of the value chain), which meant that only small-scale pilot plants came to fruition. The new administration of MAS's President Luis Arce (2021–) called on foreign companies to look into other ways of extracting lithium since they have a problem recovering lithium through brine extraction.

Mexico has also recently entered the lithium race. Following in Bolivia's footsteps, it nationalised lithium in April 2022. Mexico ranks tenth out of 23 countries with mineral reserves, with the world's largest deposit in Sonora, a state northwest of the country. After nationalisation, Mexico and Bolivia agreed to create a technical team and a scientific committee in charge of exploring international cooperation projects for lithium exploitation, production and processing.

LITHIUM AND THE LIMITS OF RENEWABLE ENERGIES

Faced with the scenario of dispossession and plunder taking shape in Latin America concerning lithium, it is worth asking what kind of energy transition we have in mind. History teaches us that there are no pure transitions and that the path will not be linear. Nor is there a manual on the subject containing questions and answers, much less on a large scale posed by the socio-ecological and climate crisis within the framework of complex socio-economic and socio-environmental systems. However, neither should we jump on the bandwagon of an unsustainable transition, such as that proposed in the Atacamenian salt flats, associated with transnational corporations, consolidating a green colonialism that reproduces domination over Nature and populations.

It is not true that every post-fossil society leads to post-development. The transition cannot be reduced solely to a change in the energy matrix, guaranteeing the continuity of an unsustainable consumption model. Decarbonisation of the economy must necessarily lead to a comprehensive change in production, consumption and distribution; it should aim to change the system of social relationships and reinforce the eco-dependent bond with Nature.

The post-fossil transition cannot be an excuse to consolidate or maintain openly unsustainable consumption models. No planet can endure or has enough lithium to meet its needs if we do not change the models we use for our mobility. It is not sufficient to merely replace fossil fuel-based cars with electric cars. What we need is to reduce consumption and move towards public and shared mobility models so that they can become sustainable. The fact that lithium batteries, as well as wind and solar projects, also require minerals (including copper, zinc and others) warns us of the need to reform the transportation system and, in general, the consumption model.

Numerous studies have emphasised that the energy transition, as proposed from a corporate perspective, is unsustainable from the metabolic point of view and implies an exacerbation of the exploitation of natural resources. For example, the report entitled 'Minerals for Climate Action: The Mineral Intensity of the Clean Energy Transition' (2020) proposes that the extraction of minerals:

> such as graphite, lithium and cobalt, could experience an increase of almost 500 percent between now and 2050 to meet the growing demand for clean energy technologies. It is estimated that more than 3 billion tonnes of minerals and metals will be required to implement wind, solar, and geothermal energy, as well as energy storage, to achieve a temperature reduction below 2C in the future.

Looking at it this way, the socio-ecological transition involves a broader outlook that should serve to ask ourselves more radical questions about the type of society in which we want to live and the development models we propose for the future. We must abandon our energy needs dependent on fossil fuels because they threaten the planet's life and are great polluters. However, the energy transition cannot lead us to look to false solutions, which continue to dispossess populations and reinforce social and territorial inequalities, further exacerbating the international division of labour that exists today. Nor can it be the excuse to consolidate and/or maintain unsustainable consumption models. The transition must be fair, both from an environmental and social point of view. This means that, as dependent and peripheral countries, we need to build a just society from different dimensions, redefining the challenges we face in terms of politics and civilisation.

In short, lithium's role in changing the system is not determined or unequivocal. These types of fears and questions point to a need to adopt a more comprehensive and multi-dimensional perspective about environmental costs, as well as territorial and social dimensions relating to the rights of the populations involved in the territories, as well as the rights of Nature and the role of the state, knowledge and scientific research.

NOTES

1. This article expands on ideas developed in a collective research book by the Group of Critical and Interdisciplinary Studies of Energy Problems, coordinated by Pablo Bertinat and Maristella Svampa. It reflects and reproduces parts of the works of Argento and Bertinat; Argento, Puente and Slipak, and finally, Svampa and Bertinat (*The Energy Transition in Argentina*, Svampa and Bertinat, Siglo XXI, Argentina, March 2022).

2. Pablo Bertinat, Jorge Chemes and Lyda Fernanda Forero. *Energy Transition: Contributions for Collective Reflection* (Transnational Institute y Taller Ecologista, 2021), https://transicion-energetica-popular.com/wp-content/uploads/2021/11/Energy-Transition-report-web.pdf.

3. Maristella Svampa. 'Imágenes del fin: Narrativas de la crisis socioecológica en el Antropoceno'. *Nueva Sociedad* 278 (2018).

4. Luis González Reyes. *Educar para la transformación ecosocial* (Madrid: Fuhem, 2018), https://repositorio.comillas.edu/xmlui/handle/11531/44358.

5. Roberto Andrés. 'Los medios latinoamericanos ven la transición energética solo desde una perspectiva económica y de negocios, según un estudio'. *El Diario Argentina*, 22 April 2022, www.eldiarioar.com/sociedad/medio-ambiente/medios-latinoamericanos-ven-transicion-energetica-perspectiva-negocios-economico-estudio_1_8926068.html.

6. Ministry of Energy and Mining of Argentina, 2017, cited by Julián Zicari, Bruno Fornillo and Martina Gamba. 'El mercado mundial del litio y el eje asiático. Dinámicas comerciales, industriales y tecnológicas', in Bruno Fornillo (ed.), *Litio en Sudámerica, geopolítica, Energía, Territor*ios (Buenos Aires: El Colectivo, 2019).

7. Zicari et al. 'El mercado', 62.

8. Melisa Argento and Florencia Puente. 'Entre el boom del litio y la defensa de la vida. Salares, agua y territorios y Comunidades en la región atacameña', in Bruno Fornillo (ed.), *Litio en Sudámerica, geopolítica, Energía, Territorios* (Buenos Aires: El Colectivo, 2019).

9. 'Cada tonelada de litio requiere la evaporación de 2 mil litros de agua'. Chilesustentable.net, May 2019, www.chilesustentable.net/cada-tonelada-de-litio-requiere-la-evaporacion-de-2-mil-litros-de-agua/ (last accessed 18 January 2020).

10. We clarify that the authors of this book participated in said tribunal. It can be viewed in its entirety at the following link: www.rightsofnaturetribunal.com/tribunal-chile-2019.

11. Rafael Poveda Bonilla. *Estudio de caso sobre la gobernanza del litio en Chile* (Santiago: CEPAL, 2020); Jeannette Sánchez, Rafael Domínguez, Mauricio León, Jose Luis Samaniego and Osvaldo Sunkel. *Recursos naturales, medio ambiente y sostenibilidad: 70 años de pensamiento de la CEPAL* (Santiago: CEPAL, 2019).

12. Albemarle. 'Albemarle y Consejo de Pueblos Atacameños exponen en Sustainable Mining 2019', last modified September 2019, www.albemarlelitio.

cl/news/albemarle-y-consejo-de-pueblos-atacameos-exponen-en-sustainable-mining-2019.

13. Medios el independiente. 'Bolivia crea la empresa estatal Yacimientos del Litio Boliviano', www.elindependiente.com.ar/pagina.php?id=163689 (last accessed 19 January 2020).

14. Bruno Fornillo et al. (eds). *Litio en Sudámerica, geopolítica, Energía, Territorios* (Buenos Aires: El Colectivo, 2019), p. 236.

3

Decolonising the Energy Transition in North Africa

Hamza Hamouchene

INTRODUCTION

The potential of the Sahara Desert in North Africa to generate large amounts of renewable energy thanks to its dry climate and vast expanses of land has long been touted. For years, Europeans, in particular, have considered it a potential source of solar energy that could satisfy a sizable chunk of European energy demands.

In 2009, the Desertec project, an ambitious initiative to power Europe from Saharan solar plants was launched by a coalition of European industrial firms and financial institutions. It was predicated on the hyperbolic idea that a tiny surface of the Sahara can meet the total electricity demand of the world. So, in the first decade of this century, an 'international' consortium of companies formed the Desertec Industrial Initiative (Dii), with weighty players such as E.ON, Munich Re, Siemens and Deutsche Bank all signing up as shareholders. It was formed as a largely German-led private-sector initiative aiming to translate the Desertec concept into a profitable business project by providing around 20 per cent of Europe's electricity by 2050 via special high voltage, direct current transmission cables.[1]

The export-oriented Desertec project should be seen within a context of pro-corporate trade deals and a scramble for influence and energy resources, reminiscent of 'colonial' schemes of appropriation and plunder. In these, the Sahara is usually described as a vast empty land, sparsely populated, representing an El Dorado of renewable energy, thus constituting a golden opportunity to provide Europe with electricity so it can continue its extravagant consumerist lifestyle and profligate energy consumption. However, this deceptive narrative obfuscates questions of ownership and sovereignty. It masks ongoing global relations of hegemony and domination that facilitate the plunder of resources, the privatisation of commons and the dispossession of communities, consolidating thus undemocratic and exclusionary ways of governing the transition.

After some years of hype around it, the Desertec venture ultimately stalled amid criticisms of its astronomical costs and neo-colonial connotations. However, the idea seems to have been granted a new lease of life as the possible answer to Europe's renewable hydrogen needs. In early 2020, Dii Desert Energy launched the MENA Hydrogen Alliance, which brings together private and public sector actors, science and academia to kick-start green hydrogen economies.[2] Before analysing this supposedly 'new' initiative, dubbed Desertec 3.0, it is worth glancing over some large North African solar projects that went ahead despite the earlier demise of the Desertec plans and learn some lessons on how the transition towards renewable energy can, in fact, enshrine dispossession and reproduce the same patterns of exploitation and plunder.

ENERGY TRANSITIONS, DISPOSSESSION AND EXPROPRIATION

Some transitions to renewable energy can be extractivist in nature and maintain the same dispossession practices, dependencies and hegemonies. A few examples from the North African region (with a focus on Morocco) come to mind. They all show how energy colonialism is reproduced in the form of green colonialism or green grabbing.

While Morocco's goal to increase the share of renewable energy in its energy mix to 52 per cent by 2030 in terms of installed capacity is laudable, a critical assessment must, however, be made if what really matters to us is not just any kind of transition but rather a 'just transition' that would benefit the impoverished and marginalised in society, instead of deepening their socio-economic exclusion.

The Ouarzazate Solar Plant was launched in 2016, just before the climate talks (COP22) held in Marrakesh. It was praised as the world's largest solar plant and the Moroccan monarchy was declared to be a champion of renewable energies. But scratching a little under the surface reveals another picture. First, the plant was installed on the land (3,000 hectares) of some Amazigh agro-pastoralist communities without their approval and consent, which constitutes a land grab for a supposedly green agenda (a green grab). Second, this mega-project is controlled by private interests and has been built through contracting a huge debt of 9 billion USD from the World Bank, European Investment Bank and others. This debt is backed by Moroccan government guarantees, which means potentially more public debts for a country already overburdened with debts. Third, the project is not as green as it claims to be. Using concentrated thermal power (CSP) necessitates extensive use of water to cool down and clean the panels. In a semi-arid region like Ouarzazate, diverting water use from drinking and agriculture is just outrageous.[3]

The 'Noor Midelt' project constitutes Phase II of Morocco's solar power plan and aims to provide more energy capacity than the Ouarzazate plant. It is a hybrid between CSP and photovoltaic (PV) solar power. With 800MW planned for its first phase, it will be one of the world's biggest solar projects to combine CSP and PV technologies. In May 2019, a consortium of EDF Renewables (France), Masdar (UAE) and Green of Africa (Moroccan conglomerate) was selected as the successful bidder to construct and operate the facility in partnership with the Moroccan Agency for Solar Energy (MASEN) for a period of 25 years. The project contracted more than 2 billion USD in debts so far from the World Bank, African Development Bank, European Investment Bank, French Development Agency and KfW.[4]

Construction on the project started in 2019, while commissioning is expected in 2024. The Noor Midelt solar complex will be developed on 4,141 hectares site on the Haute Moulouya Plateau in central Morocco, approximately 20 km northeast of Midelt town. A total of 2,714 hectares is managed as communal/collective land by the three ethnic agrarian communities of Ait Oufella, Ait Rahou Ouali and Ait Massoud Ouali. At the same time, approximately 1,427 hectares are declared as forest land and currently managed by the communities. The land has been confiscated from its owners through national laws and regulations allowing expropriation to serve the public interest. The expropriation was granted in favour of MASEN by the administrative court decision in January 2017, and the court decision was publicly disclosed in March 2017.

Reminiscent of an ongoing colonial environmental narrative that labels the lands to be expropriated as marginal and underutilised, and therefore available for investing in green energy, the World Bank, in a study conducted in 2018,[5] stressed that 'the sandy and arid terrain allows only for small scrubs to grow, and the land is not suitable for agricultural development due to lack of water'. This argument/narrative has also been used when promoting the Ouarzazate plant in the early 2010s. One person back then stated:

> The project people talk about this as a desert that is not used, but to the people here, it is not desert, it is a pasture. It is their territory and their future is in the land. When you take my land, you take my oxygen.[6]

The World Bank report does not stop there but goes on to assert that 'the land acquisition for the project will have no impacts on the livelihood of local communities'. However, the transhumant pastoralist tribe of Sidi Ayad who has been using that land to graze its animals for centuries beg to differ. Hassan El Ghazi, a young shepherd, declared in 2019 to an activist from ATTAC Morocco:

> Our profession is pastoralism, and now this project has occupied our land where we graze our sheep. They do not employ us in the project, but they employ foreigners. The land in which we live has been occupied. They are destroying the houses that we build. We are oppressed, and the Sidi Ayad region is being oppressed. Its children are oppressed, and their rights and the rights of our ancestors have been lost. We are 'illiterates' who do not know how to read and write… The children you see did not go to school and there are many others. Roads and paths are cut off… In the end, we are invisible and we do not exist for them. We demand that officials pay attention to our situation and our regions. We do not exist with such policies, and it is better to die, it is better to die![7]

In this context of dispossession, misery, under-development and social injustice, the people of Sidi Ayad have been voicing their discontent since 2017 through several protests. And on February 2019, they carried out an open sit-in, leading to the arrest of Said Oba Mimoun, member of the Union of Small Farmers and Forest Workers, and his sentencing for twelve months in jail.

Mostepha Abou Kbir, another trade unionist who has been supporting the struggle of the Sidi Ayad tribe, described how the land was enclosed without the approval of the local communities who have been enduring decades of socio-economic exclusion. In fact, it has been fenced and no-one is allowed to approach. He contrasts the mega-development projects of the Moroccan state with the inexistent basic infrastructure in Sidi Ayad. Moreover, he points to another dimension of the enclosure and resource grab, which is the exhaustion of water resources in the Drâa-Tafilalet region for the sake of these gargantuan projects (the Midelt solar plant will be fed from the nearby Hassan II dam) that communities complain they do not benefit from.[8] In the challenging context where small herd owners are being driven out of the sector while concentrating wealth in a few hands, along with the com-moditisation of the livestock market and chronic droughts, the Midelt solar project stands to exacerbate the threat to the livelihoods of these pastoralist communities and worsen their marginalisation.

It is not only Sidi Ayad communities who have been voicing concerns about this project. Some women from the Soulaliyate movement have also been demanding their right to access land in the Drâa-Tafilalet region and ordered the appropriate compensation for their ancestral land on which the solar plant has been built. The 'Soulaliyate women' refers to tribal women in Morocco who live on collective land. The Soulaliyate women's movement began in the early 2000s and arose in the context of intense commodification and privatisation of collective lands.[9] Tribal women demanded equal rights and shares when their land was privatised or divided. Despite intimidation, arrests and sieges by public authorities, the movement has become nationwide, and women from different regions have rallied behind the banner of equality and justice.

Despite all these concerns and injustice, the project is going ahead, protected by the monarchy and its repressive and propaganda tools. It seems that the logic of externalising socio-ecological costs and displacing them through space and time, characteristic of the extractivist drive of capitalism, has no end.

GREEN COLONIALISM AND OCCUPATION IN WESTERN SAHARA

While some of the projects in Morocco, like the Ouarzazate and Midelt Solar Plant, can qualify as 'green grabbing', the appropriation of land and resources for purportedly environmental ends,[10] similar renewable projects (solar and wind) that are taking place in the occupied territories of Western Sahara can be simply labelled 'green colonialism' as they are carried out in spite of the Saharawis and on their occupied land.

'Green colonialism' can be defined as the extension of the colonial relations of plunder and dispossession (as well as the dehumanisation of the other) to the green era of renewable energies, with the accompanying displacement of socio-environmental costs onto peripheral countries and communities. Basically, the same system is in place, but with a different source of energy, moving from fossil fuels to green energy, while the same global energy-intensive production and consumption patterns are maintained and the same political, economic and social structures that generate inequality, impoverishment and dispossession remain untouched.

At present, there are three operational wind farms in occupied Western Sahara. A fourth is under construction in Boujdour, while several are in the planning stage. Combined, these wind farms will have a capacity of over 1,000MW. These wind farms are part of the portfolio of Nareva, the wind energy company that belongs to the holding company of the Moroccan royal family. Ninety-five per cent of the energy that the Moroccan state-owned phosphate company OCP needs to exploit Western Sahara's non-renewable phosphate reserves in Bou Craa are made from windmills. The renewable energy is generated by 22 Siemens wind turbines at the 50MW Foum el Oued farm, operational since 2013.[11]

In November 2016 at the time of the UN Climate Talks COP22, Saudi Arabia's ACWA Power signed an agreement with MASEN to develop and operate a complex of three power stations of solar photo-voltaic (PV) totalling 170MW. Two of those power stations (operational today), totalling 100MW, are however not located in Morocco, but inside the occupied territory (El Aaiún and Boujdour). Plans have also been issued for a third solar plant at El Argoub, near Dakhla.[12] These renewable projects are being used to entrench the occupation by deepening Morocco's ties to the occupied territories, with the obvious complicity of foreign capital and companies.

After this small detour, let's now come back to Desertec and hydrogen.

HYDROGEN, THE NEW ENERGY FRONTIER IN AFRICA

Clean or green hydrogen refers to the extraction of hydrogen from more complex substances using 'clean' (zero carbon) processes. Most current hydrogen production is the result of extraction from fossil fuels, leading to large carbon emissions (grey hydrogen). Through carbon capture technology, for example, this process can be made cleaner (blue hydrogen). However, the cleanest form of hydrogen extraction uses electrolysers to split water molecules, which can be powered by electricity from renewable energy sources (clean or green hydrogen).

The EU's hydrogen Strategy published in July 2020 – in the European Green Deal (EGD) framework – is an ambitious roadmap for shifting towards green/clean hydrogen by 2050. It proposes that the EU could meet some of its future supply from Africa, particularly North Africa, which offers both huge renewable energy potential and geographic proximity.[13]

The idea originated in a paper published in March 2020 by trade body Hydrogen Europe setting out the '2 x 40 GW green hydrogen initiative'. Under this concept, by 2030, the EU would have in place 40GW of domestic renewable hydrogen electrolyser capacity and import a further 40GW from electrolysers in neighbouring areas, among them the deserts of North Africa, using existing natural gas pipelines that already connect Algeria and Libya to Europe.[14] One of the authors of this Hydrogen Europe's 2 x 40 GW initiative paper also co-wrote the Dii North Africa-Europe hydrogen 'manifesto' in November 2019.[15]

Within Europe, Germany is among those at the forefront of green hydrogen efforts in Africa. It is working with the Democratic Republic of Congo, Morocco and South Africa to develop 'decarbonised fuel' generated from renewable energy for export to Europe and is exploring other potential areas/countries particularly suited to green hydrogen production.[16] In 2020, the Moroccan government entered into a partnership with Germany to develop the first green hydrogen plant on the continent. As always, Morocco, boasting one of the region's most neoliberal(ised) economies, keeps garnering praise for its business-friendly environment, openness to foreign capital and its 'leadership' in the renewable energy sector. According to certain estimates, the country can take up to 4 per cent of the global Power-to-X market (production of green molecules) by 2030, given its 'exceptional renewable resources and its successful track record in deploying large-scale renewable plants'.[17]

However, it is important to stress from the outset that what is at stake in all these documents, manifestos, policy papers and initiatives is the EU's energy security, first and foremost. Europe, its priorities and vision are always centre stage, and all the rest need to be reshaped and forced to adapt, albeit with some rhetoric around the shared and trickle-down benefits for all the parties involved.

For instance, the Desertec proposal, which advocates for a European energy system based on 50 per cent renewable electricity and 50 per cent green hydrogen by 2050, starts from the presumption that 'due to its limited size and population density, Europe will not be able to produce all its renewable energy in Europe itself'. Therefore, it assumes that a large part of the hydrogen will be imported and what a better region than the sun and wind-endowed North Africa to secure this. The new Desertec proposal attempts to distance itself from the focus on exports from the initiative's early days, by adding the dimension of local development of a clean energy system. However, according to project proponents, the export agenda cannot be underestimated or shunned away: 'over and beyond catering for domestic demand, most North African countries have huge potential in terms of land and resources to produce green hydrogen for export'.[18] If that wasn't convincing enough for the political and business elites on both sides of the Mediterranean, the Desertec crew have other tricks up their sleeves: 'Furthermore, a joint European–North African renewable energy and hydrogen approach would create economic development, future-oriented jobs and social stability in North-African countries, potentially reducing the number of economic migrants from the region to Europe.'[19] I am not sure if that's desperate or akin to hard selling strategies, but it seems that the Desertec vision lends itself to the general mood of consolidating Fortress Europe and expanding an inhuman regime of border imperialism, while trying to tap into the low-cost energy potential of North Africa that relies on undervalued and disciplined labour.

Desertec is thus presented as one solution to Europe's energy transition, an opportunity for economic development in North Africa and a brake on South–North migration. However, if it is really serious about addressing these issues, it needs to target their structural causes that lie in a destructive and unjust global economic model. Being rather an apolitical techno-fix, it promises to overcome these problems without fundamental change, basically maintaining the status quo and the contradictions of the global system that led to these problems in the first place. In this vein, it embraces the obsession with endless economic growth, repackaged in the oxymoron 'green growth' and gives the illusion of the endless availability of energy and resources, thus indirectly perpetuating consumerist lifestyles. This will do nothing to bring our socio-economic system within the planetary boundaries, in time to avert the climate and ecological breakdown.

Big engineering-focused 'solutions' like Desertec tend to present climate change as a shared problem with no political or socio-economic context. This perspective hides the historical responsibilities of the industrialised West, the problems of the capitalist energy model, and the different vulnerabilities between countries of the North and the South. Moreover, by using language such as 'mutual cooperation', 'for the benefit of both', that presents the Euro-Med region as a unified community (we are all friends now, fighting against a common foe!), it masks the real enemy of African people, which is neocolonial structures of power, exploiting them and plundering their resources.

Furthermore, by pushing for the use of the current gas pipeline infrastructure, it effectively advocates for a mere switch of the energy source while maintaining the existing authoritarian political dynamics and leaving intact the hierarchies of the international order. It might be too much to ask this of Desertec, but the fact that it is encouraging the use of pipelines from Algeria and Libya (through Tunisia and Morocco) begs the question of what will be the future of the populations in these two fossil fuel-rich countries. What will happen when Europe stops importing gas from them (13 per cent of the gas consumed in Europe is from North Africa)? What about the ongoing chaos and destabilisation caused by the NATO intervention in Libya? Would Algerians' aspirations for democracy and sovereignty – well expressed in the 2019–21 uprising against the military dictatorship – be considered in this equation? Or is it simply another remake of the status quo where hydrogen replaces gas? Perhaps, there is nothing new under the sun after all.

To add insult to injury, the Desertec manifesto points out that 'in an initial phase (between 2030–2035), a substantial hydrogen volume can be produced by converting natural gas to hydrogen', whereby the CO_2 is stored in empty gas/oil fields (blue hydrogen). This, alongside the use of the rare water resources to produce hydrogen, can be considered as yet another example of dumping waste in the Global South and displacing environmental costs from North to South (the creation of sacrifice zones), a strategy of imperialist capital where environmental racism is wedded to energy colonialism.

Last but not least, when talking about the mutual benefits for Europe and North Africa, the manifesto says: 'But North Africa lacks the technology, capital and a well-educated labour force to develop a clean energy system on their own.' But why is this so in the first place? Doesn't this have to do with relations of ongoing domination and appropriation of wealth? Isn't this attributable to monopolising technology and the intellectual property regime that showed its cruelty in the current pandemic? Isn't it because of all the imposed structural adjustment programmes that hollowed out public services such as health and education in these countries? Nevertheless, the issue around knowledge and technology transfer is primordial. Like any other technology, the questions of who uses it, who owns it, how it is implemented, for what agenda, and in which context it is being promoted are of great importance.

And if we assume that the Desertec approach is the way forward, one challenge remains: its cost. A huge upfront investment would be needed to establish the infrastructure required to produce and transport green hydrogen. Given previous experiences carrying out such high-cost and capital-intensive projects (Ouarzazate solar plant as an example), the investment ends up being more debts for the receiving country, deepening the dependence upon multilateral lending and foreign assistance.

CONCLUSION

In such context, it is fundamentally important to scratch beneath the surface of the language of 'cleanliness', 'shininess' and 'carbon emission cuts' to observe and scrutinise the materiality of the transition towards renewable energy. The analysis attempted to examine different projects and initiatives through the lens of creating new commodity chains, revealing that effects can be no different from the destructive fossil fuel and mining activities in the region and beyond.

What seems to unite all the aforementioned projects and the hype around them is a deeply erroneous assumption that any move toward renewable energy is to be welcomed and that any shift from fossil fuels, regardless of how it is carried out, is worthwhile. One needs to say it clearly: the climate crisis we are currently facing is not attributable to fossil fuels per se but rather to their unsustainable and destructive use to fuel the capitalist machine. In other words, capitalism is the culprit. If we are serious in our endeavours to tackle the climate crisis (only one facet of the multi-dimensional crisis of capitalism), we cannot elude questions of radically changing our ways of producing and distributing things, our consumption patterns and fundamental issues of equity and justice. It follows from this that a mere shift from fossil fuels to renewable energy sources while remaining in the capitalist framework of commodifying and privatising Nature for the profits of the few will not solve the problem. If we continue down this path, we will only end up exacerbating or creating another set of problems around land ownership and natural resources issues.

Most writing on sustainability, energy transitions and environmental issues in North Africa is dominated by international neoliberal institutions and think tanks. Their analysis is limited and does not include class, race, gender, power or colonial history questions. The solutions and prescriptions proposed by them are market-based and take a top-down approach without addressing the root causes of the climate and ecological crises. The 'knowledge' produced by such institutions is profoundly disempowering and overlooks questions of oppression and resistance, focusing largely on the advice of 'experts' to the exclusion of voices 'from below'. In all cases, ordinary people and the working poor are excluded from any strategy and painted as inefficient, backward and unreasonable. The North Africans whose lives will be affected the most by the climate/ecological crisis (and the top-down and unjust ways of addressing it) will be the small family farmers, fisher-folk, pastoralists (whose rangelands are being appropriated to build mega-solar plants and wind projects), workers in the fossil fuel and extractive industries, informal workers and the pauperised classes. But they are sidelined and prevented from shaping their future. Instead, economic, developmental and energy policies are often shaped by entrenched domestic and international interests.

A green and just transition must fundamentally transform and decolonise our global economic system, which is not fit for purpose at the social, ecological and even biological levels (as revealed by the COVID-19 pandemic). It also necessitates an overhauling of the production and consumption patterns that are energy-intensive and utterly wasteful, especially in the Global North. In this respect, degrowth can be an avenue to explore the cores of the global system.

We must always ask the relevant as ever questions: who owns what? Who does what? Who gets what? Who wins, and who loses? And whose interests are being served? We need to break away from the imperial and racialised (as well as gendered) logic of externalising costs that, if left unchallenged, would only generate green colonialism and further pursuit of extractivism and exploitation (of Nature and labour) for a supposedly green agenda.

The fight for climate justice and a just transition needs to consider the differential responsibilities and vulnerabilities between North and South. So ecological and climate debt must be paid to countries in the Global South that are the hardest hit by global warming and have been locked by global capitalism in a predatory extractivism.

In a global context of forced liberalisation, the push for unjust trade deals, an imperial scramble for influence and energy resources, green transition and talk about sustainability must not become a shiny façade for neocolonial schemes of plunder and domination.

NOTES

1. Hamza Hamouchene. 'Desertec: the renewable energy grab'. *New Internationalist*, 1 March 2015,. https://newint.org/features/2015/03/01/desertec-long (last accessed November 2022).
2. Dii Desert Energy. 'MENA Hydrogen Alliance', https://dii-desertenergy.org/mena-hydrogen-alliance/ (last accessed April 2023).
3. Hamza Hamouchene. 'The Ouarzazate solar plant in Morocco: triumphal "green" capitalism and the privatization of nature'. *Jadaliyya*, 2016. Available at: www.jadaliyya.com/Details/33115.
4. NS Energy. 'Noor Midelt Solar Power Project, Morocco', www.nsenergybusiness. com/projects/noor-midelt-solar-power-project-morocco/ (last accessed May 2023).
5. The World Bank. Report No: PAD2642. Project paper on a proposed additional loan in the amount of US$ 100 million and a proposed clean technology fund loan in the amount of US$ 25 million to the Moroccan Agency for Sustainable

Energy (MASEN) for a Morocco Noor solar power project additional financing. 15 May 2018, https://documents1.worldbank.org/curated/en/1384815286878 21561/pdf/Morocco-Noor-AF-project-paper-P164288-May17-clean-05212018. pdf.

6. Karen Rignall. 'Theorizing sovereignty in empty land: contested global landscapes'. Land Deal Politics Initiative, October 2012, www.yumpu.com/en/ document/view/35781099/theorizing-sovereignty-in-empty-land-contested-global-landscapes.

7. ATTAC Morocco. *Oh Land*, produced in 2019, documentary, 16:00, https://m. facebook.com/attac.maroc/videos/199096351435545/.

8. Ibid.

9. ATTAC Morocco. 'The Soulaliyate movement: Moroccan women fighting land dispossession', 2020. Available at: https://waronwant.org/news-analysis/ soulaliyate-movement-moroccan-women-fighting-land-dispossession.

10. James Fairhead, Melissa Leach, and Ian Scoones. 'Green grabbing: a new appropriation of nature?' *Journal of Peasant Studies* 39(2) (2012): 237–61.

11. Western Sahara Resource Watch. 'Dirty green energy on occupied land', https:// wsrw.org/en/news/renewable-energy> (last accessed April 2023).

12. Ibid.

13. European Commission. 'A Hydrogen Strategy for a Climate-Neutral Europe', 2020, https://ec.europa.eu/energy/sites/ener/files/hydrogen_strategy.pdf (last accessed April 2023).

14. John Parnell. 'European Union sets gigawatt-scale targets for green hydrogen', www.greentechmedia.com/articles/read/eu-sets-green-hydrogen-targets-now-blue-hydrogen-has-to-keep-up (last accessed April 2023).

15. Dii Desert Energy. 'A North Africa–Europe Hydrogen Manifesto', 2019, https:// dii-desertenergy.org/ (last accessed April 2023).

16. Clifford Chance. 'Focus on hydrogen: a new energy frontier for Africa', 12 January 2021, www.cliffordchance.com/briefings/2021/01/focus-on-hydrogen--a-new-energy-frontier-for-africa.html (last accessed May 2023).

17. Fraunhofer Institute for Systems and Innovation Research ISI. 'Study on the opportunities of "POWER-TO-X" in Morocco – 10 hypotheses for discussion', February 2019, www.econbiz.de/Record/study-on-the-opportunities-of-power-to-x-in-morocco-10-hypotheses-for-discussion-eichhammer-wolfgang/ 10012238280.

18. Dii Desert Energy, 'Manifesto'.

19. Ibid.

4

Can the Greatest Polluters Save the Planet? Decarbonisation Policies in the US, EU and China[1]

John Feffer and Edgardo Lander

The United States, the European Union and China are responsible for the largest share of carbon emissions in the world today: a total of 52 per cent with China contributing more than half.[2] They are also responsible for more than half of all emissions historically, with the United States and the EU having emitted 47 per cent and China about 13 per cent.

These numbers obscure, however, what has happened since 1990, when governments first began to discuss the importance of reducing carbon emissions. From 1990 to 2017, global emissions increased by 63 per cent.[3] The EU, over that same period, slashed its emissions by about 20 per cent. By contrast, the United States managed a reduction of a mere 0.4 per cent, while China's emissions increased by 350 per cent. European countries took the risk of climate change very seriously, the United States went back and forth on the issue depending on the politics of whatever administration was in power, and China continued to prioritise economic growth, arguing its 'right to develop'.

The legacy of the last 30 years still weighs heavily on the environmental and energy policies of these three key actors. Today, on the surface, all three have acknowledged the importance of rapidly scaling back on carbon emissions as part of an unprecedented economic transformation. The United States and EU have promised carbon neutrality by 2050, while China has pledged to achieve peak emissions by 2030 before reaching carbon neutrality by 2060. But the EU is moving forward quickly, the United States fitfully, and China not yet.

The approaches of these three world powers to decarbonisation also reflect their respective political economies. The EU with its Green Deal has forged a compromise among its members that pairs a 'clean' industrial strategy with various market mechanisms, which mirrors the EU's social democratic roots alongside its more recent adoption of neoliberal programs. Individual member states, meanwhile, have adopted more stringent decarbonisation strategies that could set important global precedents.

The United States has adopted a piecemeal approach at the federal level that scatters various 'clean energy' incentives across a range of policies under a series of misleading names – the Inflation Reduction Act, the Bipartisan Infrastructure Law, the CHIPS and Science Act – while certain states like California and cities like Ithaca, New York have taken much bolder actions. Vested interests, particularly fossil fuel companies, have exerted considerable political influence to undermine any efforts to coordinate a more effective federal approach.[4]

China plans to continue to prioritise economic expansion, putting off overall cuts in carbon emissions until after 2030. It continues to rely heavily on fossil fuels to power this expanding economy, with an overwhelming reliance on coal, the most polluting of these energy sources. However, China has also pushed for a rapid expansion of renewable energy, with special emphasis on wind and solar. It is adding more such capacity than any country in the world.[5] Here, too, some provinces are pushing for more rapid transformation and this could lead to an earlier peak in emissions for the country as a whole and an earlier deadline for carbon neutrality.

In this chapter, we will evaluate the 'green deals' of the United States, the EU and China to see how far they fall short of their own rhetoric and the carbon reductions necessary to keep global temperatures below the 1.5 degree increase from pre-industrial levels by 2050 established by the Paris Agreement. We will also look at how these transformations are contingent on a variety of mechanisms that shift carbon-intensive agriculture, manufacturing and services to the countries of the Southern hemisphere as a strategy to externalise emissions. At the heart of the transitions in the world's most polluting countries is a persistent zero-sum mentality that reduces carbon emissions and the negative impacts of extractivism in the North largely by exporting those problems to the South. At the same time, the wealthiest countries refuse to address the underlying driver of climate change: overconsumption.

THE EUROPEAN UNION'S GREEN DEAL

On paper, European countries are taking climate change very seriously. Finland has pledged to be carbon neutral by 2035, Austria by 2040 and both Germany and Sweden by 2045. A number of European countries – Denmark, France, Hungary – have even put their commitments into law. These commitments are even more serious because European countries have some of the worst carbon footprints in the world. In terms of per-capita emissions, Germany is number seven at 10.4 metric tons per person while France clocks in at number 14 with 6.6 tons (which is also roughly the EU average).[6]

These carbon neutrality pledges are all within the realm of the possible. Unlike other parts of the world, European countries are acting in concert in response to climate change. In July 2021, the European Union unveiled its 'Fit for 55' plan by which the regional bloc of 27 countries aims to reduce its collective emissions by 55 per cent by 2030. This carbon reduction plan is part of a larger 'European Green Deal' first introduced in December 2019 that promises 'economic growth decoupled from resource use'.[7] This larger plan, which European member states are still debating, envisions increasing the share of renewables to 40 per cent of overall energy use, renovating 35 million buildings to make them more energy-efficient, while creating 160,000 new green jobs in the construction sector, and boosting organic farming as part of a 'Farm to Fork' program that aims to make agricultural production, distribution and consumption more sustainable.[8]

All of this will, of course, cost a lot of money. The EU has pledged to devote as much as 30 per cent of its long-term budget, which would amount to around US$700 billion, to climate action.[9] As part of the revenue collection side of the plan, a Carbon Border Adjustment Mechanism (CBAM) would effectively apply a tariff on carbon-intensive goods coming into the EU. A Just Transition Mechanism of around US$85 billion over six years would help poorer regions of the bloc meet the plan's goals. Within this mechanism, a 'public sector loan facility' would combine grants from the EU budget with financing from the European Investment Bank. The European Union also issued its first 'green bond' in the expectation that it would bring in US$14 billion for its Green Deal budget.

Integral to paying for the European Green Deal is a heavy reliance on private finance as well as modifications to the EU's existing Emissions Trading Scheme (ETS). Established in 2005 and currently the world's largest carbon market, the ETS covers factories, power stations and the airline industry. The Fit for 55 plan proposes to include emissions from ships and, in a separate new market, road transport and the building sector. The higher price for carbon that will likely result has divided the EU along its east–west axis, particularly since Eastern Europe is more heavily dependent on fossil fuel than the West.

European policymakers have repeatedly acknowledged the scale of the climate crisis and the urgency of acting sooner rather than later. But however ambitious the European Green Deal might look on paper, it remains insufficient. Implementing this landmark initiative might well be a victory, even an impressive victory when compared to what China or the United States is doing. But the European Green Deal does not fully live up to its name. Its job creation promises are rather anaemic. Compare, for instance, the opening up of a mere 160,000 new green jobs in infrastructure to the 400,000 EU workers directly involved in the oil, gas and coal sector in 2016, a number that has held steady (with the exception of coal mining) for some years.[10]

At first glance, Europe is putting together a considerable amount of money for its Green Deal. The Sustainable Europe Investment plan expects to mobilise US$1 trillion by 2030. Around half this money will come directly from the EU budget, which will in turn trigger national co-financing of US$114 billion over the period 2021 to 2027. A guarantee fund called Invest EU will also spur public and private investment of US$279 billion. And the Just Transition Mechanism will assist poorer and more fossil fuel-dependent areas to keep pace with the rest of the EU.

But a lot of this money is just old wine in new bottles. The InvestEU Fund, for instance, is simply a continuation of the older European Fund for Strategic Investment. At the same time, the EU continues to use up a lot of the remaining global carbon budget through more loopholes and offsets in carbon markets while continuing to pour billions of dollars into fossil fuel subsidies and expanding fossil fuel infrastructure within Europe and in Africa in the wake of Russia's invasion of Ukraine.[11]

Also, while the EU talks about setting up a fund to tackle inequality *within* Europe, it has done relatively little to address global inequality. Little money has been set aside to help Europe's trading partners meet the new stringent requirements of the CBAM. According to IMF consultants He Xiaobei, Zhai Fan and Ma Jun, 'CBAM could result in an annual welfare *gain* in developed countries of US$141 billion, while developing countries see an annual welfare *loss* of US$106 billion, compared to a baseline scenario.'[12]

However, individual European countries have provided down payments on both climate financing and the loss and damage currently experienced in the Global South. Of the US$100 billion annually that richer countries promised to transfer to poorer countries for adaptation and mitigation, several European countries – Germany, Norway, Italy, Sweden – have worked hard to meet their obligations, though the overall global figure has officially reached about US$83 billion, much of this money comes in the form of new loans and insurance.[13] On loss and damage, Scotland at the 2022 COP in Sharm el-Sheikh pledged an additional £3 million on top of what it offered the previous year, bringing its total to £5 million. This commitment prompted Ireland, Austria, Belgium, Denmark and Germany to follow suit. In the end, 200 nations at the COP 27 in Sharm el-Sheikh agreed to a new loss-and-damage fund, though it remains unclear how much money this new facility will deliver to the hardest hit countries of the South, and how.

THE US GREEN NEW DEAL

In November 2018, the Green New Deal (GND) became a rallying cry for US climate activists when members of the Sunrise Movement occupied House Speaker Nancy Pelosi's office and adopted the slogan as their unifying message. A few months later, Rep. Alexandria Ocasio-Cortez (D-NY), who had joined the young activists in Pelosi's office, brought this message to Congress when she partnered with Sen. Ed Markey (D-MA) to introduce their Green New Deal resolution.

More manifesto than binding legislation, the resolution laid out a vision of an equitable clean energy transition for the United States. It was full of bullet points and broad proposals, such as 'invest in the infrastructure and industry of the United States to sustainably meet the challenges of the 21st century' and provide all Americans 'with access to clean water, clean air, healthy and affordable food, and nature'.

In drawing from the language and history of Franklin D. Roosevelt's New Deal of the 1930s, climate activists hoped to bring together two strands of the progressive movement: environmentalism and economic justice. Activists urged the United States to radically reduce its carbon footprint and, at the same time, create well-paying jobs, especially for those workers leaving economic sectors associated with dirty energy. As with Roosevelt's program, the Green New Deal relies on government direction and funding to advance this major economic transformation.

Since the original resolution, other Green New Deal bills have emerged, on education, housing and cities. US cities, too, have established Green New Deal initiatives at municipal level, and many civic organisations continue to champion the GND as a radical vision for a reoriented US society. In 2022, many earlier climate-related provisions appeared in a single legislative package. The Inflation Reduction Act, framed as an effort to address rising prices, reduce federal debt and provide targeted economic assistance, contains a raft of clean-energy provisions from climate justice block grants to the creation of a national green bank. These provisions add up to about US$369 billion in new spending, the largest federal investment into clean energy in US history. It also puts the United States much closer to achieving the Biden administration's goal of halving US carbon emissions by 2030.

In terms of specific investments, the Act offers a kind of industrial policy for the United States by funnelling US$60 billion into 'clean energy' manufacturing and job creation around solar panels, batteries and other components. This comes on top of as much as US$67 billion in investments in zero-carbon industries and climate research in the CHIPS Act. As part of the Biden administrations Justice 40 approach[14] – in which 40 per cent of all climate spending is supposed to go to disadvantaged communities – the Act directs money to low-income households to electrify their homes. Allocations also include funds to electrify public transportation and Postal Service vehicles, building on the cleaner transportation investments in the Bipartisan Infrastructure Law. Another US$20 billion will go to farmers to switch to sustainable practices like crop rotation. The bill for these programs will largely be offset by higher taxes, including a 15 per cent minimum tax on corporations with over US$1 billion in revenue and a one percent excise tax on corporate share buybacks.

Unfortunately, the Act also makes some disheartening concessions to fossil fuel companies, including the expansion of mining and drilling permits and a tax credit that could keep coal-fired plants in operation. Some of the funding, particularly the CHIPS Act, frames 'clean energy' investments as integral to a more aggressive policy toward China, in geopolitical terms.

US states remain divided on these environmental and energy issues, with some still heavily invested in fossil fuels (coal in West Virginia, oil in Texas, natural gas in Pennsylvania). These divisions make a more coordinated federal policy difficult to achieve. It also complicates any national commitment to international efforts. The Paris Agreement obligations are voluntary, for instance, principally because US negotiator John Kerry made clear that a deal with mandatory targets wouldn't pass Congress. Moreover, while the EU and China have had largely consistent political positions across different administrations, the United States has swung wildly from cooperation

(Obama) to hostility (Trump) and back again to cooperation (Biden). Such political mood swings undermine any US efforts to be a leader on this issue.

CHINA'S GREEN REVOLUTION

When it comes to a global clean energy transition, China is both part of the problem and part of the solution. On the problem side, China is the largest emitter of carbon dioxide in the world by a rather wide margin (though it is only number four in terms of per-capita emissions).

At the same time, China has been a global leader in shifting from fossil fuels to renewable energy, adding more renewable energy capacity than any other country. By the end of 2022, China is on pace to install 156 gigawatts of additional capacity provided by wind turbines and solar panels, which is 25 per cent more than the record it set in 2021.[15] By comparison, the United States is expected to install only about 30 gigawatts of solar and wind power this year.[16]

China's economy continues to grow, albeit less dramatically than in previous decades, and so do its energy needs. Total power usage has increased about 4 per cent so far in 2022 compared to 2021.[17] Since China made its first international pledges to tackle climate change in 2009, its economy has grown threefold – but its energy consumption has only grown by half that figure. China has also been a driver of international climate agreements. Its 2014 bilateral climate deal with the United States made possible the subsequent Paris climate agreement.

In the space of one generation, China transformed itself into a global economic giant. It now faces a task of comparable urgency and scale. In the space of a generation, China must lead the world by greening its enormous economy. How quickly Beijing can and will accomplish this goal will largely determine whether the world can prevent the global temperature from exceeding 1.5 degrees Celsius over pre-industrial levels.

Despite its commitment to expand its renewal energy infrastructure, China remains the leading consumer of fossil fuels in the world, using twice as much as the United States. Moreover, more than half of China's energy consumption comes from coal, which releases more carbon into the atmosphere than oil or natural gas.

Yet the Chinese government has pledged to reach peak carbon dioxide emissions before 2030 and to achieve carbon neutrality by 2060. The timeline for China is highly compressed, trying to achieve in a single generation what took Europe or the United States two generations. The share of coal is still about 56 per cent in primary energy consumption, which it needs to reduce sharply and quickly.[18] China has also pledged to bring the share of non-fossil energy sources – wind, solar, hydro, biomass and nuclear – to 80 per cent of total energy consumption by 2060.[19] It might be able to achieve this goal sooner if climate-forward provinces end up driving national policy.

Both the United States and China have devoted considerable energy to establishing standards that can raise the environmental standards of development projects. The United States has been instrumental in establishing the Blue Dot Network, which promotes 'quality infrastructure investment that is open and inclusive, transparent, economically viable, Paris Agreement aligned, financially, environmentally and socially sustainable, and compliant with international standards, laws and regulations'. China, meanwhile, has developed a 'traffic light system' to ensure that its Belt and Road Initiative (BRI) projects reduce environmental risks and contribute to a green transformation, with green representing a positive contribution, yellow neutral, and red negative.

It's a matter of debate whether China is working fast enough to shift to clean energy. What is not debatable, however, is China's predictability. It has established goals and followed through on achieving them. What makes Chinese positions reasonably consistent over time is the current leadership's determination to increase the country's energy security by reducing dependency on external suppliers of fossil fuels. The government in Beijing is also acutely aware of public support for cleaner air, land and water, which has generated protests in the past that have challenged regime stability. A third motivating factor is China's desire to position itself as a global climate leader at a time when US climate policy has seesawed wildly. China also has a certain credibility as a 'late developer' that is only now reconsidering its dependency on fossil fuels, which can be persuasive to countries in the South trying to balance development and decarbonisation.

THE NEW GREEN COLONIALISM

Although the Global North is rushing to embrace new 'clean' technologies like wind and solar, it has largely been adding renewables without reducing fossil fuel consumption or scaling back on energy use. Overconsumption in the North continues to reinforce a neocolonial power imbalance with the South.

Consider, for instance, the cornerstones of the 'clean transitions' in the North: solar cells, windmill turbines and lithium-ion batteries for electric cars. First, Green (New) Deals and similar initiatives continue to put growth at the centre of economic policy – and with that growth, high energy consumption. Second, these technological improvements are designed to maintain a high-consumption lifestyle that is linked to significant carbon emissions. Third, these innovations require significant inputs from the Global South: copper and aluminium for solar panels, zinc, rare earth elements and balsa wood for wind turbines, and lithium, cobalt and nickel for battery storage. Finally, the intended use of these technologies often shows an individualist bias that omits the necessary structural transformation: electric cars for individual use or decarbonisation campaigns targeted at individual consumers (for instance, to reduce the use of plastic bags or plastic straws).

The North has also developed new conceptual frameworks to promote the 'clean energy transitions'. To promulgate the idea that decarbonisation can take place at current levels of consumption in richer countries, the earlier campaign for 'sustainable development' has been hooked up to the more contemporary agenda of 'renewable energy'. But many of the components of this renewable energy – lithium, cobalt – are just as finite as oil and gas, which calls into question the sustainability of the enterprise. The blind spot on these inputs is comparable to the externalisation of environmental costs that has long accompanied conventional measures of economic growth (see also Bengi Akbulut's chapter on feminist degrowth in this volume). In other words, if the true cost of pollution to land, water and air were factored into many manufacturing processes, the latter would not in fact be profitable; similarly, if the true environmental costs of the inputs were factored into 'clean energy' technologies, the latter would not in fact be sustainable.

Another conceptual framework is the primary focus on decarbonisation to the exclusion of other pressing environmental and economic concerns. This framework begins with the 'carbon footprint', which was originally the brainchild of an advertising firm contracted by British Petroleum, which puts the onus on the individual consumer rather than the institutional contributors to climate change, notably fossil fuel companies. This decarbonisation framework extends to the compacts at the heart of the Paris climate deal, which mobilise collective action to reduce carbon emissions, and even more so to carbon markets, which allow for carbon emissions in one place as long as they are compensated by reductions elsewhere – ending up in no absolute reduction at all.

While decarbonisation is essential, it is not the only environmental crisis facing the planet, which includes, among others, deforestation, biodiversity loss, the loss of soil fertility and declining access to clean water. That these latter crises are particularly acute in the South only contributes to the belief that the richest countries are focused primarily on decarbonisation because climate change threatens their specific economic interests in a way that, for instance, lack of access to clean water in the South does not. Of course, many of the 'clean energy' technologies require increased, rather than decreased, environmental damage in the South by impacting watersheds, forests and agricultural land despoiled by mining, otherwise productive land given over to large solar arrays, or the widespread logging of balsa wood to supply Chinese wind power projects.

In short, the 'clean energy transitions' of the United States, Europe and China must be evaluated not only according to the gap between pledges and global targets and the gap between stated policies and actual implementation, but also in terms of the overall net harm to the environment and peoples of the South when the full social and environmental costs to the supplying nations are factored into the equation. Under the old colonial models, the wealth and security of the North depended on the wealth plundered from the South. Under the new green colonialism, the North continues to assume that Nature and cheap labour from the countries of the South are naturally available to it in order to maintain the unsustainable and effectively imperial modes of living of its inhabitants.[20] Any Global Green New Deal must not only be fully global and equitable but also transcend the fundamental assumptions about growth and consumption that have generated the planetary environmental crises to begin with.

NOTES

1. This essay draws on four articles by John Feffer, 'What Remains of the U.S. Green New Deal?'; 'The Green New Deal Goes Local'; 'The Future of China's Green Revolution'; and 'European Green Deal: Step Forward, Backward, or Sideways?' (all published by Foreign Policy In Focus) as well as Edgardo Lander. 'La transición energética corporativa-colonial', paper given at the Transiciones justas en América Latina y el Caribe, 29–30 September 2022 (Bogota, Colombia). All hyperlink footnotes below were last accessed on 13 December 2022.

2. Hannah Ritchie and Max Roser. 'CO2 emissions'. *Our World in Data*, https://ourworldindata.org/co2-emissions.

3. 'List of countries by carbon dioxide emissions', *Wikipedia*, https://en.wikipedia.org/wiki/List_of_countries_by_carbon_dioxide_emissions.

4. Ben Lefevre. 'Easter eggs in climate bill delight oil and gas industry', www.politico. com/news/2022/07/28/manchin-oil-gas-biden-schumer-climate-bill-00048514.
5. Gavin Maguire. 'China on track to hit new clean & dirty power records in 2022.' *Reuters*, 23 November 2022, www.reuters.com/business/energy/ china-track-hit-new-clean-dirty-power-records-2022-maguire-2022-11-23/.
6. Willem Roper. 'Wealthy nations lead per-capita emissions.' *Statista*, 1 March 2021, www.statista.com/chart/24306/carbon-emissions-per-capita-by-country/; 'Greenhouse gas emission statistics - carbon footprints', *Eurostat*, March 2022, https://ec.europa.eu/eurostat/statistics-explained/index.php?title= Greenhouse_gas_emission_statistics_-_carbon_footprints.
7. European Commission. 'A European Green Deal', https://commission.europa. eu/strategy-and-policy/priorities-2019-2024/european-green-deal_en.
8. European Commission. 'Delivering the Green Deal', https://commission. europa.eu/strategy-and-policy/priorities-2019-2024/european-green-deal/ delivering-european-green-deal_en; 'Organic Action Plan', https://agriculture. ec.europa.eu/farming/organic-farming/organic-action-plan_en; 'Farm to Fork Strategy', https://food.ec.europa.eu/horizontal-topics/farm-fork-strategy_en.
9. BBC. 'Climate change: EU to cut CO2 emissions by 55% by 2030', 21 April 2021, www.bbc.com/news/world-europe-56828383.
10. Veronika Chako. 'Employment in the energy sector', European Commission, 9 July 2020.
11. Krista Larson. 'Europe turns to Africa in bid to replace Russian natural gas.' *Associated Press*, 12 October 2022, https://apnews.com/article/russia-ukraine-middle-east-africa-business-senegal-52c9da7d4d79d99fef1e35d0430 dba25.
12. He Xiaobei, Zhai Fan and Ma Jun. 'The global impact of the Carbon Border Adjustment Mechanism'. Task Force on Climate, Development, and the International Monetary Fund, 11 March 2022, www.bu.edu/gdp/files/2022/03/ TF-WP-001-FIN.pdf.
13. Tracy Carty and Jan Kowalzig. 'Climate finance short-changed', Oxfam, 19 November 2022, https://policy-practice.oxfam.org/resources/climate-finance-short-changed-the-real-value-of-the-100-billion-commitment-in-2-621426/.
14. Justice 40. www.thejustice40.com/ (last accessed 13 December 2022).
15. Stefan Ellerbeck. 'These regions produce a lot of carbon emissions – here's what they plan to do about it.' World Economic Forum, 26 August 2022, www. weforum.org/agenda/2022/08/electricity-capacity-power-renewable-energy/.
16. Benjamin Storrow. 'Wind and solar are booming, but emissions aren't falling.' *ClimateWire*, 14 October 2022, www.eenews.net/articles/wind-and-solar-are-booming-but-emissions-arent-falling/.
17. China National Energy Administration, 13 October 2022, www.nea.gov. cn/2022-10/13/c_1310669666.htm.
18. China Power, 'How is China's energy footprint changing?', https://chinapower. csis.org/energy-footprint/.
19. Ivy Yin. 'China commits to 80% of energy mix from non-fossil fuels by 2060'. *S & P Global*, 25 October 2021, www.spglobal.com/commodityinsights/en/ market-insights/latest-news/energy-transition/102521-china-commits-to-

80-of-energy-mix-from-non-fossil-fuels-by-2060#:~:text=China%20will%20
work%20toward%20having,the%20country's%20highest%20executive%20
body%2C.

20. Ulrich Brand and Markus Wissen. *The Imperial Mode of Living: Everyday Life and the Ecological Crisis of Capitalism* (London: Verso Books, 2021).

5

Accumulation and Dispossession by Decarbonisation

Ivonne Yanez and Camila Moreno

In recent decades, capitalism has reinvented itself. On the one hand, through the discourse of *sustainable development* and its institutional ramifications, which include the United Nations Framework Convention on Climate Change (UNFCCC) and its protocols, agreements and legal frameworks, or the Convention on Biological Diversity and its obsessive advances to place a third of the planet under conservation regimes. But also, on the other, through various ways of expanding capital towards new frontiers, with unimaginable merchandise and markets, for which more forms of territorial control and green masks are required. This is necessarily accompanied by confusing propaganda language, albeit one that is successful in selling and deceiving.

The decarbonisation horizon today articulates the various areas of the new green capitalism, which include energy transition, conservation and restoration mechanisms via markets, as well as new digital assets. While these three spaces may be related to each other, they do not imply a provincialisation of the economy. In fact, they lead to land control and, in many cases, to the violation of rights and a prolongation of neocolonialism. The objective of this chapter is to analyse the new faces of green capitalism, framed within the umbrella of decarbonisation, which allow the main polluters to reinvent themselves by appealing to a supposedly ecological rhetoric. Let's start with an account of the proposals for the decarbonisation of the economy and the energy transition, which are capitalist and colonial and imply an imperialist imposition, as we will see throughout the text.

THE CAPITALIST AND COLONIAL DECARBONISATION

In a scenario of extreme climatic disasters, the narrative and agenda that is imposed from the North, and which already has many followers among the hegemonic actors of the Global South, tells us that to avoid these catastrophes, we must move towards the 'energy transition' and 'decarbonise' our economies. In practice, we are facing a new type of greening of capitalism, with negative consequences for peoples and Nature.

The 'decarbonisation' proposal has its origin in green capitalism associated with the climate crisis. In 2007, the Intergovernmental Panel on Climate Change (IPCC)[1] already included this term in one of its reports on climate change mitigation and defined it as the 'way towards a low-carbon-intensity economy'. They also postulated that decarbonisation could be achieved through geoengineering plans such as bioenergy with carbon capture and storage (BECCS). These proposals from the IPCC should not surprise us since it is usually aligned with false solutions to climate change. Even the concept of *mitigation* is part of the pool of proposals that are framed in the carbon market economy, as part of the UNFCCC.

The hegemonic postulates of *decarbonise while creating new accumulation dynamics* range from the carbon offset certificate market to investment in speculative financial assets such as green or blue bonds. They also include the massive and dangerous deployment of millions of devices with technologies based on renewable energies, the expansive electrification of transportation, or the extraction of what are known as strategic minerals for the energy-digital transition, which include rare earths or lithium.

These 'decarbonisation' proposals (which do not focus on fossil carbon, but on the carbon of CO_2 molecules) are actions that allow for continuation of the model of capital accumulation and economic growth based on fossil fuels to continue. In reality, they do not represent real forms of decarbonisation (leaving barrels of oil, cubic meters of gas, or tons of coal in the ground) since, in practice, they mean more carbon and more CO_2 in the atmosphere. This is a transition in which industry, working alongside the regulatory role of the states, establishes new sources of energy such as hydrogen, while exhausted oil wells are used as opportunities for CO_2 capture and storage, tree monocultures expand for biomass and crops for agrofuels, and the narrative is expanded to relaunch nuclear energy in small modular reactors, for local and decentralised use.

In this logic of 'decarbonisation of the economy' nothing is actually decarbonised. In the search for a 'Net Zero' result for emissions, false equivalences are validated between biological carbon molecules – which are part of the life cycle – and those that flow into the atmosphere from the human action of extracting matter geological fossil (oil, gas and coal).[2] Taken together, decarbonisation actions under the 'net zero' discourse maintain the same pattern of civilisation, anchored in historical logic and mechanisms that deepen inequalities. As a consequence, this hegemonic 'decarbonisation' worsens the climate and violates human and Nature rights.

In addition, capitalist and neoliberal 'decarbonisation' ideas go hand in hand with other more recent ideas that include carbon neutrality, net zero carbon, the circular economy, agriculture 4.0 proposals, climate-smart mining, digitisation of the economy, the uncoupling of economy and Nature and digital carbon. All these plans encourage expansion of the oil frontier and a huge grabbing of millions of hectares of land as *carbon sinks* for the extraction of more minerals. Later, we will see how this also occurs in the hyper-digitisation of economic processes and digitised information that is becoming the most important asset in the global economy, under the new concept of 'digital development'.

DIRECTIONS OF THE NEW CLIMATE GOVERNANCE

In the context of the new climate governance, the twin ideas of 'decarbonisation' and 'Net Zero' have become absolutely crucial to understanding how big polluters reinvent themselves. 'Decarbonisation' is used more and more to refer to and give meaning to a shared future and a common historical horizon. The global climate governance regime that has been negotiated since 1992 with the Earth Summit in Rio de Janeiro and the rise of 'climate' as a central issue on the international environmental agenda coincide with the accelerated globalisation that took place with the post-Cold War neoliberal world order.

Since the Paris Agreement was reached in 2015, we have had a convergence of the climate and development agendas, as well as the climate financing and development agendas, within the reference framework of the United Nations 2030 Agenda for Sustainable Development (SDGs). In fact, institutions, actors and mainstream economic thinking have fully embraced the green economy paradigm set forth in the 2006 Stern Report on the economics of climate change, which consists of internalising the costs of environmental destruction, incorporating natural capital into the system of national accounts (the whole spectrum of 'environmental services') and thus make Nature visible to capital.

In this process, the international climate policy and its mechanisms have become the key transmission current to understanding how this meta-process is being incorporated into national and local contexts and how new legislative frameworks are established (such as Forest Codes, Mining Codes, water regulation, etc.) to accommodate the green economic paradigm. This also serves as a basis for building the legal foundation and legal certainty for new contracts and transactions related to natural capital and the advance of commodification on the frontier of intangibles.

How is the paradigm of the green economy and its mechanisms reflected within the territories? They require new forms of expropriation and settling in specific territories (forests, mangroves, pastures, etc.). It also involves delving into our mental structures to define the language of how we communicate about climate change and how we make sense of this shared effort. For example, the concepts of 'decarbonisation' and 'Net Zero' are already in widespread use. Increasingly, we use them to make sense of shared goals, transparency and accountability.

The political mandate to 'decarbonise' the economy and society – and now to digitise them – has become the backbone of the global climate governance regime and the long-term horizon towards climate neutrality. The hegemonic Green New Deals and many of the *just transition* options also run through this logic of 'decarbonisation'. However, if we take into account the evolution of the multilateral regime of the last three decades to address climate change within the framework of the UN, there is no agreed definition of what decarbonisation means.

WHAT DOES IT MEAN TO 'DECARBONISE'?
WHAT DOES 'NET ZERO' MEAN?

In the evolution of the regime, from the UNFCCC (1992) to the Paris Agreement (2015), we have agreement and explicit language on increasing global temperatures from pre-industrial levels. However, with the Paris Agreement, language was included relating to 'emissions based on sources and absorption through sinks',[3] incorporating the functioning of the CO_2 cycle as a field of climate action and, as such, of the global norms and mechanisms of climate governance. From this perspective, we have seen the initially confusing concept of 'Net Zero' rapidly gain ground, reaching its peak in the run-up to COP26 in Glasgow in November 2021. There, in the midst of a post-COVID multilateral agenda and framed by the narrative of Green New Deals, the 'Net Zero' slogan gained a very prominent place in the UNFCCC and in building alliances with corporations and the financial sector, including insurance. Initiatives characteristic of *multistakeholder* capitalism, such as the 'Net Zero Banking Alliance: industry-led, UN-convened' took command of the climate action umbrella.[4] In this scenario, the 'Net Zero by 2050' of the 'We Mean Business Coalition' is also a key concept.[5]

Under the UN regime's rules, it is today impossible to address what has been known as climate action separately from 'Net Zero'. Through scientific arguments, the economics of offsetting carbon emissions is justified. Market mechanisms and approaches are present, even if it is not explicitly called 'trade', for example, when we discuss the idea of Internationally Transferable Mitigation Results (ITMO). In this global accounting of emissions mitigation, they appropriate Nature and its ecological cycles that are invariably occurring in territories, inseparable from their social context and framed in political contexts as power relations.

The European Union (EU) played a crucial role in this change from the objective of reducing emissions to talking about degrees of temperature. Through the European program for adaptation and mitigation, hundreds of scientists were involved on the continent, resulting in the creation of the Representative Concentration Pathways (RCP), which became the central facet of the IPCC and of the climate negotiations from that moment onwards. The RCPs took the place of the Special Report of Emissions Scenarios (SRES), which focused mainly on social changes to face global warming, forcing a migration of the focus from emissions to degrees of temperature. The models on which the RCPs are based are complicated physical and economic evaluations, emptied of any societal overtones.

Decarbonisation through Net Zero is therefore framed in this new context of understanding climate change as molecules and formulas in the hands of scientists and not in profound transformations, including policies, to move towards a post-extractivist path. This explains why it's crucial to understand what Net Zero implies by mid-century and what this seemingly technical jargon really consists of.

TECHNICAL JARGON AND COMMERCIAL LANGUAGE

What appears under the highly technical language of the Rule Book for the operationalisation of Article 6 of the Paris Agreement,[6] the driving force of the Agreement is emblematic of what has been identified and criticised in recent decades as the new green face of capitalism, reshaping its dynamic of expropriation to extract every last drop of hydrocarbons and profit from environmental collapse.

Almost three decades have passed since the idea of carbon trading was incorporated into the global climate governance regime. With the Kyoto

Protocol (1997), what were known as flexibility mechanisms were included to allow industrialised countries (included in Annex I of the Protocol) to comply with their obligations. One of the facilities, the Clean Development Mechanism (CDM), allowed industrialised countries to carry out their emission reduction projects in 'host' countries of the Global South to create opportunities for cooperation for both parties and for the climate.

Initially, it was about the consecration of an idea in the multilateral order that came from US national environmental policy relative to the trade with particulate pollution permits. In other words, this consisted of the government granting rights to the private sector so that it could pollute up to a certain threshold. Those rights were then traded within this sector, with the objective that market forces and the rationality of profitability would support compliance with environmental regulations. This is where the notions of 'ecosystem services' and 'environmental services' come from. Under this logic, Net Zero emissions does not actually mean zero emissions, but rather that provides the option to continue polluting while getting others to ensure that the same amount of tons of carbon will be absorbed in a carbon 'sink'. This means that the emission of CO_2 in one context is equated with supposed absorption in another, without considering the conditions, the actors, the places and the specific power relations involved.

If we look at the long road that has been traversed since the idea of carbon trading first emerged, we see a massive advance of a market-based environmentalism in public policy at all levels, taking hold as the culturally hegemonic mindset on how to act ecologically: defining actions for the planet as a great business opportunity.

In their contribution to this book, John Feffer and Edgardo Lander have analysed the decarbonisation plans of the countries that are the planet's main polluters: the United States, European Union and China. The authors also show how these policies are based on a new green colonialism, which is expressed precisely in a zero-sum mentality that seeks to reduce the negative impacts of extractivism in the North, exporting problems to the South. In Ecuador, the balsa wood case is widely known,[7] since this is a product that has been in high demand in recent years, mainly by China, to build wind turbine propellers. But beyond the specific impacts on the territories, it is also important that we look at how national governments in the world's periphery are repositioning themselves in the face of these global disputes.

In Ecuador, the government is putting together the National Decarbonisation Plan through the Ministry of Environment, Water and Energy Transition, which has set up a public–private partnership with funds that include the Sustainable Environmental Investment Fund (FIAS) whose objective is, according to its website, 'to support the financing of environmental management, through the implementation of strategies and financial mechanisms for the protection, conservation, and improvement of natural resources'. Extractive oil and mining sectors, large industrialists, or agro-export companies from the banana, shrimp and palm oil sectors are also clamouring for decarbonisation.

At the same time, the Ecuadorian government announced its interest in doubling oil extraction,[8] and is expanding large-scale mining[9] and the agro-industrial frontier. Business decarbonisation also reaches cement companies, which already have the Carbon Neutral seal, for recycling garbage at their facilities, or having emission compensation certificates. For example, UNACEM, owner of the Selva Alegre cement company, received the Carbon Footprint Quantification Distinction from the Ecuador Zero Carbon Program (PECC), for contributing to the reduction of greenhouse gases. To do this, all the company needed to do was quantify and verify its gas inventory with the certifier. An example of greenwashing, this seal will allow the company to say that it is on a decarbonisation path, even though the global cement industry is responsible for at least 8 per cent of global emissions. 'If the cement industry were a country, it would be the third largest emitter, after the United States and China.'[10]

OTHER WAYS TO DECARBONISE, OTHER TIMES FOR THE TRANSITION

Decarbonising in a real and effective way requires breaking with the hegemonic discourse and with the idea of framing climate based on the Paris Agreement. In reality, it is a carbon trade agreement, at the heart of which are carbon offsets that, as has been pointed out, do not reduce emissions, but are, in practice, permits to pollute.

At the end of the 1990s and the beginning of the twenty-first century, social struggles opposed neoliberal globalisation, free trade, the subordination of nation states, and the power of the World Trade Organisation. Two decades later, the horizon and array of social struggles have given new meaning to the whole matter. The environmental issue and the turn towards the territories have gained space and prominence at the same time that green capitalism was being established as a new hegemonic agenda, especially after the financial crisis of 2008. New goods and new markets emerged in a process that crystallised from 2012 with the Rio+20 Conference and the

SDGs. The 2015 Paris Agreement is part of this new stage of the global environmental governance regime. Setting the goals of the Paris Agreement as a prospect means supporting this agenda of false solutions and the continuous carbonisation of the planet with more fossil fuels.

We need to decarbonise our language and eliminate CO_2 from being a central reference point in our discourse. This is a condition to face climate change, move towards global justice, both with people and Nature, something today sacrificed in the hegemonic energy transition. We have to rethink what it is we want to change, how, among whom, when and where.

For the Andean Kichwa peoples of Ecuador, the *transition* could be called *tukuna*, which means to transform and become, but also to become or, even more, to be able and capable of something. This requires us to rethink of temporality. It does not mean moving towards something better *in the future*, but rather, as several indigenous groups in Abya Yala propose, it is part of walking with the past ahead. We know that time is neither unique, linear or homogeneous. The key here is to think about these multiple notions of time, overlapping and in conflict, when setting out the times for the *transition*.

Waorani, Kichwa[11] or Shuar[12] times come into contradiction with modern times, which always seek to own, buy, sell and appropriate the time of workers, the body of women and the cycles and functions of Nature. Constant acceleration in time, in transportation, and in ultra-fast computers is part of the planet's destruction. In contrast, a *transition* in *Pacha* can only be completed considering time and space, the cosmos where the reproduction of life occurs, in correspondence, complementarity and reciprocity; where human persons are not the centre, but a part of the Great House.

However, the energy transition in the form of real decarbonisation must also take space into account; for example, that occupied by fossil fuels. When oil is extracted, it leaves one space for another, and at the same time occupies another temporality. This is because underground fossil carbon is not the same as carbon in trees or soils. If we do not understand the space-time element, we will not be able to build other forms of transit. What the Yasuní proposal[13] in Ecuador sought was exactly that, to move towards a truly decarbonised economy.

The nearly three decades of COPs, since 1995 in Berlin, have not only been useless in combating climate change, but they have also actually exacerbated the problem. This is the road that has led to *failed decarbonisation*. However, in this same period of time, numerous groups of people have managed to partially put a stop to oil extraction, namely the Ogoni in Nigeria, or the Kichwa People of Sarayaku in Ecuador (see the chapter by

Tatiana Roa Avendaño and Pablo Bertinat in this book). These experiences are much more inspiring for an ecological transition than what thousands of official delegates at climate summits have achieved. However, these people are also increasingly under surveillance and siege by the new digital strategies of the green economy.

DIGITAL CONTROL: NEW THREATS TO TERRITORIES AND RIGHTS

At the end of 2022, social media[14] reported on an agreement signed between an indigenous organisation from the Ecuadorian Amazon, the FICSH (Interprovincial Federation of Shuar Centres) of Ecuador, and a company called ONE AMAZON.[15] This agreement has warned about the new type of business based on the digitisation of information obtained from lands and territories of indigenous peoples. The case says a lot about where capital is moving in a context of an increasingly diversified green economy.

These agreements are usually presented as proposals to develop activities related to the protection, defence and conservation of the forests, but in general they create a system to collect all possible information from the territories and financialise the forest and its conservation. Data is collected through satellite images and other technologies (digital, documentary, videographic, auditory and of another nature), which are associated with the issuance of 'digital assets' called 'values' (tokens or Digital Security Assets) with the use of blockchain technology.

A 'digital asset' can refer to a variety of digital commodities, such as non-fungible tokens (known as NFTs) and cryptocurrencies. It is very likely that the contracts and companies that are popping up in the Amazon are interested in doing business with the information that they will obtain from indigenous territories. People anywhere in the world who buy NFTs or tokens of images associated with indigenous territories digitised and placed on the blockchain will not directly own the pieces of territory, but they could easily access the data files on the specific hectare of territory to which the NFT is associated. In other words, buyers of NFTs issued based on indigenous Amazonian territories could say that they own digital assets that can be 'authentically' linked to every living tree or insect in one hectare of the forest.

In Ecuador, a company called Bit-CO2 also operates[16] in communities of the Achuar Amazonian nationality. Bit-CO2 has issued a token of alleged 'sponsorship' of carbon conservation activities in Achuar territory[17] from

Ecuador, but, unlike NFTs, it is a cryptocurrency. In other countries, such as Brazil, the Nemus corporation issued forest NFTs,[18] which are tokens made from the ancestral forest territory of the Apurinã people. To do this, Nemus has divided the indigenous territory into small territorial squares, each potentially represented by a unique NFT that is sold on the international market. When entering the Nemus website, you can see the Apurinã indigenous territory gridded to issue NFTs.

These are examples that show the mentality of new companies that profit from NFTs, cryptocurrencies or move the Web 3.o. One effect of this is that, due to their characteristics, NFTs are ideal for quickly inflating speculative bubbles and are even fertile ground for illegal money laundering, especially those linked to drug trafficking. It is also an indirect way of land grabbing in countries of the South.

These types of companies could also be interested in building the 'Internet of Forests' (modelled on the so-called 'Internet of Things') in the Amazon. This means the implementation of technological systems to collect images and other files from the forests to obtain information, complemented by the installation of sensors, radars or other equipment in the field. Once all this infrastructure is installed and the Forest Internet has been built, it will be easy for companies to monetise this data, for example, sell them to parties who buy environmental services. Such data in and of itself would not amount to tokens of environmental services, but, in a separate process, they could be used by companies that issue offset certificates as tokens to then obtain some type of property rights over the biological capacities, cycles and functions of Nature in the Amazon. It is important to remember that the tokens of environmental services are routinely sold to companies and states as permits to destroy Nature, biodiversity and water sources elsewhere on the planet. They are also sold as rights to pollute the atmosphere with carbon dioxide or other greenhouse gases.

In an economy in which digitisation increasingly prevails, the Amazon forests are being transformed into a Big Data bank and also related to the sale of *tokenised* environmental services. This may mean that possible environmental services tokens will be equivalent to destruction rights purchased by public or private institutions and would have the potential to harm lives and territories. There is also the risk that the information collected by the infrastructure that will collect data for companies in indigenous territories could be sold to states or private companies that need information to carry out synthetic biology, or even for debt swap plans. In practical terms, the collection and sale of relevant data on indigenous territories could indirectly reduce the access, control and self-government capacity of indigenous peoples over their territory.

All these examples are disastrous manifestations – in this case, digital – of hegemonic decarbonisation, in which large corporations and financial capital become partners, with the support of many states. While violating rights, it also promotes a climate, technological and financial structural adjustment program that legalises an entire data architecture for the production of false equivalences, violent 'offsets' and new subalternities.

NOTES

1. 'Fourth Assessment Report: Climate Change', Working Group III: Mitigation of Climate Change, Intergovernmental Panel on Climate Change (IPCC), 2007: 219, www.ipcc.ch/report/ar4/wg3/.
2. Acción Ecológica. 'Glosario de la Justicia Climática', 16 June 2022, www.accionecologica.org/glosario-de-la-justicia-climatica/.
3. Paris Agreement. United Nations, 2015, Art. 35, https://unfccc.int/sites/default/files/english_paris_agreement.pdf.
4. UN Environment Programme. 'Net-Zero Banking Alliance', www.unepfi.org/net-zero-banking/ (last accessed May 2023).
5. We Mean Business Coalition. 'We are in a decisive decade', www.wemeanbusinesscoalition.org/ (last accessed May 2023).
6. Article 6 of the Paris Agreement, concluded in Glasgow, is the carbon markets rule book. It opens the door to voluntary negotiation on emission reductions between countries, also known as 'mitigation results', to achieve their objectives proposed in the national documents (NDC) under the supervision of the UNFCCC. It also acknowledges 'non-commercial' approaches and introduces financing and technology transfer. Until today, the negotiations on the implementation roadmap of this article still continue.
7. Ivonne Yánez and Elizabeth Bravo. 'La Energía Eólica. El Case de China', in *Energías Renovables, Selvas Vaciadas. Expansión de la energía eólica en China y la tala de balsa en el Ecuador* (Quito: Naturaleza de Derechos, 2021), www.naturalezaconderechos.org/wp-content/uploads/2021/09/LA-BALSA-SE-VA.pdf.
8. Diego Cazar Baquero. 'Duplicar la producción petrolera, la controvertida apuesta del gobierno de Ecuador', Plan V, 1 September 2021, www.planv.com.ec/historias/sociedad/duplicar-la-produccion-petrolera-la-controvertida-apuesta-del-gobierno-ecuador.
9. Empresa Nacional Minera. 'Plan de Negocios 2020', 2020, www.enamiep.gob.ec/wp-content/uploads/2020/04/PLAN-DE-NEGOCIOS-EXPANSI%C3%93N-E-INVERSI%C3%93N-2020.pdf.
10. Jonathan Watts. 'Concrete: the most destructive material on Earth', *The Guardian*, 25 February 2019, www.theguardian.com/cities/2019/feb/25/concrete-the-most-destructive-material-on-earth.

11. The Amazonian Kichwas have a 'way of living in the present without talking about the future', which becomes a counter-meaning of modern times and the imposition of planning policies focused on progress where the past disappears. Meanwhile, the Waorani understand 'before' and 'after' in a different way.

12. Acción Ecológica. 'Amazenas del Capitalismo Digital: el caso de One Amazon', January 2023, www.accionecologica.org/observaciones-al-convenio-de-asociacion-entre-one-amazon-y-la-ficsh/.

13. The proposal seeks to conserve the region's biodiversity, protect the indigenous peoples living in voluntary isolation inside the Yasuní National Park and mitigate global warming by leaving the oil in the soil. On 20 August 2023, Ecuador voted to halt all future oil drilling in the Yasuní-ITT and close all oil activities in this block.

14. Poder Shuar's Profile. Facebook, 24 November 2022, www.facebook.com/photo/?fbid=435368702143515&set=pcb.435368718810180.

15. Acción Ecológica. 'Amazenas del Capitalismo Digital'.

16. See Bit-CO2. 'Ecosistema Descentralizado de Compensación Ecológica', www.bit-co2.net/es_index-tokens; Green CryptoCurrency Chanel, Youtube, 2020, www.youtube.com/@greencc9274/discussion.

17. Bit-CO2. 'Información del Proyecto', https://bit-co2.net/es_index-proj-add/oc6fny6; Bit-CO2. 'Proyectos de Conservación Forestal / Nodo-08', www.bit-co2.net/es_index-proj-card-achuo1.

18. Nemus is a collectible NFT designed to protect the Amazon Rainforest. Guardian App, https://app.nemus.earth/map.

PART II

Analysing Green Colonialism: Global Interdependencies and Entanglements

6

The Continuity and Intensification of Imperial Appropriation in the Global Economy

Christian Dorninger

INTRODUCTION

Environmental historians have demonstrated that the rise of Western Europe is closely related to the coercive appropriation of natural resources and labour during the colonial period. Later, European settler colonies in North America or Australia also developed a similar imperial orientation towards the Global South. The widespread narrative that with the end of colonial rule, imperial forms of appropriation have also come to an end, has been under critique at least since the 1950s.[1] However, imperial appropriation is much more subtle today and camouflaged under free trade and market price rules premised on the understanding that such relations are economically beneficial to all participating parties.[2] It should be noted that the unequal exchange between nation-states is preceded by 'internal colonialism', an unequal appropriation by core-like areas within nation-states,[3] which can, however, not be grasped with a country-level analysis as the present study.

The theory of ecologically unequal exchange postulates, in contrast to mainstream economics, that the *materiality* of trade and economic production and consumption are key to understanding prolonged inequalities and interdependencies between richer and poorer regions in the world. Asymmetric net transfers of resources from poorer to richer world regions, including the materials, energy, land and labour embodied[4] in all kinds of traded goods and services, underlie the current international trade regime. This unequal appropriation of resources – the theory of ecologically unequal exchange goes – has a self-reinforcing character as net-appropriating countries are then able to generate more and higher value-added goods and services, allowing them to act as net appropriators of resources in following years without having to experience commensurate socio-environmental impacts from resource extraction.[5] Adopting a biophysical perspective in assessing economic activities has the potential to shed light on pressing issues of global justice (the distribution of environmental goods and burdens) amid efforts for sustainability transformation.

The Global North depends on resource-intensive industrial technologies and infrastructures whose seamless functioning is contingent on annual net inflows of resources and embodied labour from distant and often poorer areas of the Global South.[6] Moreover, countries of the Global North obtain significantly higher monetary compensation for the resources and embodied labour they export than poorer nations, which often correlates with the positions they occupy in global supply chains.[7] As measured in asymmetric net transfers of resource volumes and monetary transfers, this inequality is crucial for individual countries to achieve economic growth and accumulate capital and technological infrastructure.[8]

This chapter strongly builds on recent publications on the global patterns of ecologically unequal exchange[9] and the drain from the Global South through imperial appropriation.[10] These contributions trace how patterns of ecologically unequal exchange and imperial appropriation did not stop with the end of colonial rule but are finding continuation to date. Here, I briefly revise these findings with a new database and for more recent years.

METHODS AND DATA

For this chapter, I used Release 055 of the GLORIA global environmentally extended multi-region input-output (MRIO) database,[11] constructed in the Global MRIO Lab.[12] Environmentally extended input-output analyses allow calculating consumption-based accounts that are often referred to as footprint indicators. The footprint indicator of a given country for each socio-environmental indicator is calculated as the sum of the domestic extraction/use of the resource plus the upstream resource uses embodied in the country's imports, less the direct and indirect resource requirements to produce goods and services for exports. For this chapter, I use the following environmental extensions: raw material extraction (measured in metric tons), land use (hectares), energy supply (joule) and labour requirements (person-year equivalents).[13]

PATTERNS OF ECOLOGICALLY UNEQUAL EXCHANGE

In a first step, in Figure 6.1, I assessed the exchange of three natural resources plus labour relevant to the economic production of goods and services, i.e., raw material equivalents, energy, land and labour that are embodied in a countries' exports, to determine the ecologically unequal exchange between the Global North and the Global South over the last three decades.[14] The 'Global North' is represented by the IMF's 'advanced economies' and involves the current EU countries (except for Bulgaria, Croatia, Hungary, Poland and Romania), Australia, Canada, Hong Kong, Iceland, Israel, Japan, New Zealand, Norway, Singapore, South Korea, Switzerland, the United Kingdom and the United States of America. The larger rest of the countries belong to the IMF's 'emerging and developing economies' and represent the

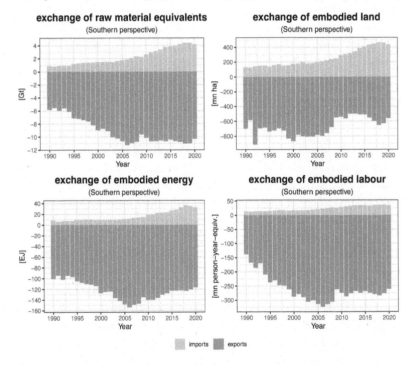

Figure 6.1 The exports and imports of raw material equivalents (measured in gigatons), embodied land (in million hectares), embodied energy (in exajoules), and embodied labour (in million person-year-equivalents) between the Global North and the Global South. Bars below the zero line indicate the direct and indirect export of resources and labour, bars above the zero-line imports. Net exports from the South to the North are calculated as exports minus imports.

'Global South' in this study. It should be noted that proceeding with this country grouping locates China in the Global South, which might be disputable for particularly China having gained significant weight during these three decades both in biophysical and geopolitical terms. However, this shift can only be given a cursory treatment in this chapter. Figure 6.4 provides an analysis of monetary drain from the Global South in relation to their GDP – once with and once without China.

Figure 6.1 shows a relatively stable net appropriation of the Global North from the Global South across all four flows assessed and the observed period. Northern net appropriation is defined as the exports from the South to the North less Southern imports from the North. For most indicators, some sort of stabilisation of ecologically unequal exchange ratios materialised after the global financial crisis of the year 2008. Results indicated that for raw material equivalents, the Global South net provided around 7 gigatons (exports minus imports) per year from 1990 to 2020 to the Global North. In terms of primary energy embodied in internationally traded goods and services, the Global North net appropriated around 107 exajoules per year and almost 240 million person-year equivalents over the same period. The South also acted as a constant net provider of embodied land for Northern final consumption, at around 430 million hectares per year, but for the ten most recent years (i.e., from 2011 to 2020), this average net provision decreased to around 160 million hectares per year, indicated by fewer exports of embodied land since 2007 until 2015 and steadily growing imports until 2018. In sum, the economies of the Global North rely on a constant net-inflow of natural resources and labour from the Global South which allows economic growth while conserving domestic resources.

MONETARY REPRESENTATION OF UNEQUAL EXCHANGE

In the following, I aim to represent this unequal exchange in terms of value using the TiVA (Trade in Value Added) indicator in constant 2015 USD.[15] However, it is utterly important to understand what can be expressed in terms of monetary exchange values and what not. First, there is no such thing as a 'correct or fair price' for any good or service that would, in any sense, compensate the ecologically unequal exchange analysed above. Second, monetary value can also not be used to define what the global South *could* earn under fairer conditions, because the world trade scheme would look very different without these price differentials in place. So, this exercise is not about finding an exchange price at which international trade between the Global North and South would become fair, but rather to simply represent the drain from the South in terms of existing market prices and thus to uncover the advantage for the Global North caused by these unequal exchange structures including price differentials.

DRAIN REPRESENTED IN NORTHERN PRICES

To begin with, I will represent the monetary drain from the South in terms of Northern prices, i.e., the monetary compensation that countries of the Global North receive per unit of embodied resource flow, as suggested by Amin[16] and Köhler.[17] To assess drain (as the value transfer through unequal exchange) from the South in monetary terms, I multiply the net resource drain from the South (as shown in the previous section) with the compensation Northern countries receive for the same unit of exports ('Northern prices'), and subtract the actual net monetary transfer from the North to the South. To maximise the comparability of resources and TiVA, I only used internationally traded flows, excluding domestic-only flows. The following equation can be used to represent the drain from the South valued in Northern prices:

$$T = R_{net} * P_N - M_{net}{}^{18}$$

The left graph of Figure 6.2 shows the results of this equation when applied to the four indicators assessed. It shows that land use, followed by mate-

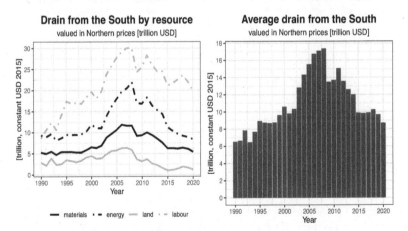

Figure 6.2   Drain from the South represented in Northern prices, trillion constant 2015 USD, 1990–2020. The graph on the left shows the drain based on each of the four indicators assessed; the right graph shows the average from these four flows.

rials, is the embodied resource flow most equally compensated. The two most unequally compensated are embodied energy and labour. When averaging the drain based on these four types of flows (right graph), we can see that the average drain valued in Northern prices increased from around US$6 trillion in 1990 to more than 17 trillion USD in 2008, after which it decreased again to more than US$8 trillion in 2020. Over the 1990–2020 period, this annual average drain sums up to US$343 trillion and represents a significant windfall for the Northern economies.

DRAIN REPRESENTED IN GLOBAL PRICES

For comparison, I here assess the Southern drain valued in global prices, that is, the monetary compensation for exports (TiVA) by any country (Northern and Southern countries). In doing so, I evaluate both the drain due to the Northern price deviation from this global average price and the drain due to the Southern price deviation from the global average price. The following equation is the basis for this calculation:

$$T = R_{net} * P_G - M_{net} \text{ [19]}$$

On the left graph of Figure 6.3, the drain valued in global prices is distinguished between the four parameters assessed and shows again that labour

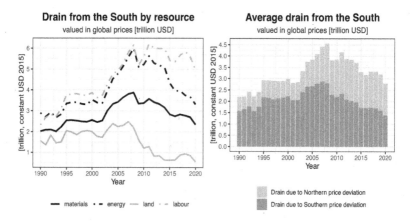

Figure 6.3   Drain from the South represented in global prices, trillion constant 2015 USD, 1990–2020. The graph on the left shows the drain based on each of the four indicators assessed; the right-side graph shows the average from these four flows differentiated by drain due to Northern price deviation and drain due to Southern price deviation.

and energy are the most unequally compensated inputs, while materials and particularly land are more equally compensated, resulting in a lower drain. Averaging these four types of drain, in the right panel of Figure 6.3, the light grey represents the share of drain that is due to the deviation of Northern prices above the global average, while the dark grey represents the share that is due to the deviation of Southern prices below the global average. Summing up, over the three decades, the annual drain expressed in global prices represented a total benefit of around US$101 trillion for the Global North.

DRAIN IN RELATION TO GROSS DOMESTIC PRODUCTS (GDP)

Having assessed the drain from the South both in Northern prices as well as in global prices, to evaluate its significance for the respective economies, one can compare the results (1) to the Northern GDP in terms of Northern prices to estimate the resulting advantage for Northern economies, in contrast to a counterfactual scenario where the production of the respective goods would have taken place in Northern countries instead of outsourcing it to Southern countries; and (2) to the drain measured in global prices to the Southern GDP, to illustrate the Southern losses compared to an equal trading world.

Doing so, in the top graph of Figure 6.4 we can see that the ratios have been significant, particularly for the North. On average, drain from the Global South measured in Northern prices amounted to 26 per cent of Northern GDP over the 1990–2020 period. This represents a substantial windfall and cost-saving exercise for the Global North. For the Global South, the drain as measured in global prices amounted to 23 per cent of Southern GDP on an annual average rate over the past three decades. However, the ratio decreased from over 30 per cent at the turn of the century to below 10 per cent by 2020. This is because the Southern GDP increased sharply, especially from 2003 to 2013, mostly due to China. However, the price differentials (or, put differently: the terms of trade) between the Global North and the Global South did not converge.

To disentangle the effects that were only due to exchange with China and its GDP development, the bottom graph of Figure 6.4 shows the drain as percentage of Northern and Southern GDP excluding all exchanges with China and also excluding the Chinese GDP from the 'Southern GDP'. Looking at the Northern gains as percentage of Northern GDP first, we can see that the drain from the South decreased when excluding the unequal exchange with China by an average of 5 per cent over the observed period (compared to the graph on the top). However, excluding China's unequal exchange but also its

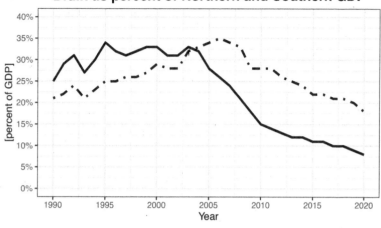

Drain as percent of Northern and Southern GDP

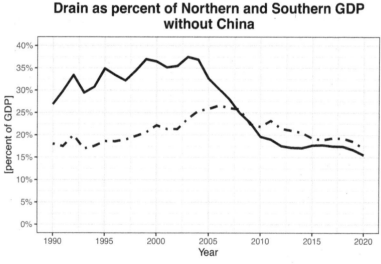

Drain as percent of Northern and Southern GDP without China

Figure 6.4 Southern drain measured in Northern prices as a percentage of Northern GDP and measured in global prices (top), and the same drain measurement excluding China from the South (bottom).

GDP from the Global South, results in an increase of drain from the South as percentage of Southern GDP by 4 per cent on average between 1990 and 2020. Meaning that an exclusion of China from the Global South increases drain as percentage of Southern GDP to an average of 27 per cent over the observed period.

Still, the Global North also gained from the drain of China, but the peak was in 2006, when the difference between Northern gains in percentage of Northern GDP including unequal exchange with China versus excluding China was more than 8 per cent (comparing the two dotted lines on the top and bottom graphs). This ratio decreased to 1 per cent in the years of 2019 and 2020.

China's GDP increased by more than 18 times from 1990 to 2020 and made up only 9 per cent of the total Southern GDP in 1990, but almost 44 per cent in 2020. However, excluding this dramatic increase in Chinese economic production from the GDP of the Global South (bottom graph), we can still see a steep decrease of Southern losses as percentage of Southern GDP between 2003 and 2013. This decrease is due to GDP increase in other Southern countries because the drain represented in global prices did not decrease to such an extent (compare right graph of Figure 6.3).

DRAIN IN RELATION TO OFFICIAL DEVELOPMENT ASSISTANCE (ODA)

To get the scale of the drain from the South into another perspective, further challenging the narrative of the Global North aiming to help 'develop' the Global South, I here compare the drain resulting from unequal exchange with the flows of official development assistance (ODA). For this analysis, a new country list has been formed which distinguishes between 'donors' (i.e., the Development Assistance Committee, DAC) and 'receivers' of ODA.

The results indicate that the amount of ODA appears extremely small relative to the drain resulting from the unequal exchange. While the drain measured in Northern prices averaged US$ 9.5 trillion per year from 1990 to 2020, and in global prices US$3 trillion, ODA merely amounted to an annual average of 0.11 trillion USD. That is, the drain in Northern prices outstrips ODA by a factor of 86, and the drain expressed in global prices exceeds ODA by a factor of 27. It seems that reducing unequal exchange provides more leverage for creating an 'even playing field' than focusing on increased ODA.

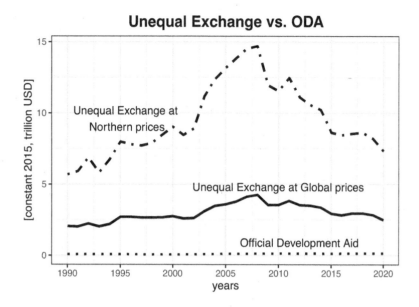

Figure 6.5 Drain from the South resulting from unequal exchange expressed in Northern prices and global prices compared to the official development assistance (ODA), from 1990 to 2020, in trillion constant 2015 USD.

DISCUSSION AND CONCLUSION

I started the empirical analysis in this chapter with assessments of contemporary ecologically unequal exchange at the global scale. The analysis reveals how high-mass consumption and economic growth in the countries of the Global North are fuelled by the maintenance of asymmetric exchange of resources and labour time with the Global South. This net appropriation is found for all three resource indicators and labour assessed and applies to the whole assessed period. This continuous reallocation of environmental goods and burdens in the international division of labour and Nature underscores the resource dependency of high-income economies on poorer economies. But this ecologically unequal exchange also implies that people and Nature in the Global South suffer disproportionally from the environmental damage inflicted by resource exploitation directly and indirectly serving richer countries. According to Hickel et al.,[20] this 'quantity of Southern raw materials, land, energy and labour *could* be used to provision for human needs and develop sovereign industrial capacity in the South, but instead it is mobilized around servicing consumption in the global North'.

The fifth indicator assessed, the TiVA to represent monetary compensation at prevailing market prices, not only shows that resources and labour embodied in exports of the Global South are compensated for at significantly lower levels than compared to Northern exports, but also that this unequal access to productive resources allows total higher TiVA in Northern countries, and that is higher GDP rates, i.e., economic growth. This also highlights that growth is fundamentally a matter of appropriation and that the apparently neutral advancement of technological innovation in the North relies on differences in how natural resources and human labour time are compensated in different parts of the world.[21]

The TiVA indicator was used to represent these price differentials as well as the monetary drain for Southern countries and, at the same time, the monetary benefit for the Global North. Even though the replications of Hickel et al.'s [22] analyses with the new GLORIA database in this chapter show partly deviations due to a different set of countries and updates in the environmental extensions of the new database, the overall patterns of ecologically unequal exchange and monetary drain remain the same.

Measuring unequal exchange in terms of Northern prices, that is, assuming the countries of the Global South would be compensated with the same amount as Northern exports per exported unit of embodied resource or labour input, the drain – as an average of the four assessed indicators – from the Global South is extremely high and sums up to more than 340 trillion USD over 31 years (1990–2020). This corresponds to around 26 per cent of the GDP generated in the Global North during the same time period. So, the drain from the South translates into a direct ecological and economic benefit for the North.

Expressing this unequal exchange in terms of global prices, i.e., if Northern and Southern countries' exports would be compensated with the annual global average prevailing market price (per unit of embodied resource or labour), the Southern drain would be lower – since global prices are much lower than Northern prices. However, hypothesising a global average price would also imply reduced compensation for Northern exports. Drain measured in global prices amounts to 101 trillion USD from 1990–2020, which corresponds, on average, to 23 per cent of Southern GDP during the same period (excluding China, this percentage increases to 27 per cent).

The analysis also indicates that the four indicators assessed – embodied materials, energy, land and labour – are compensated very differently. Energy and labour inputs seem to bear the highest inequalities in compensation. For example, in 2020, labour inputs embodied in exports from countries of the

Global North received around eight times more monetary compensation than the same unit of labour embodied in exports of the Global South. In fact, the argument that differences in compensation for labour input are due to higher productivity in countries of the Global North is truncated. Rather, labour productivity itself is a function of previous resource appropriation which enable industrialisation, the accumulation of industrial-technological infrastructure and the emergence of service sectors in the first place. Moreover, 'wage inequalities exist not because Southern workers are less productive but because they are more intensively exploited'.[23] Also compare Suwandi et al. (2019)[24] for a deeper analysis of this subject.

However, what can further be detected from most assessments in this chapter, is that a downswing in the intensity of unequal exchange occurred for some indicators starting after the year 2008 (compare Figures 6.2 and 6.3). This is likely due to the aftermaths of the financial crisis in 2008 but also to growing South–South trade and growing Southern economies in general. However, what should be noted is that particularly the 'Global South' as defined in this chapter, is highly heterogeneous. It involves countries like China as well as all least-developed countries (I have excluded China from one analysis presented in Figure 6.4 to hint at this issue).

This chapter also challenges mainstream visions of sustainable development and development assistance in general and of international trade in particular. Official development assistance appears to be not more than a drop in the ocean when compared to the structural drain experienced by the Global South on an annual basis. Considering the biophysical aspects of international trade and economic activity also suggests that a 'catch-up' development is structurally impossible in most parts of the world, as the high-mass consumption in Northern countries relies on outsourcing and net-appropriation of resources from global markets, which means continuous net-inflows of resources and labour from poorer regions.

To this end, pathways of transformation toward sustainability should not only take into account the global natural limits our economies are operating in, but also the structures and magnitude of unequal exchange, in order to move towards global justice and a dignified life for all.

NOTES

1. Raúl Prebisch. *The Economic Development of Latin America and Its Principal Problems* (New York: Lake Success, 1950); Walter Rodney. *How Europe Underdeveloped Africa* (London: Bogle-L'Ouverture Publications, 1972); Immanuel Wallerstein. *The Modern World-System* (New York: Academic Press, 1974).

2. Robert Feenstra. *Advanced International Trade: Theory and Evidence* (Princeton, NJ: Princeton University Press, 2015).

3. Joan Martinez-Alier, Leah Temper and Federico Demaria. *Social Metabolism and Environmental Conflicts in India* (Delhi: Springer India, 2016), pp. 19-49.

4. Embodied refers to the resources that not necessarily traded and crossing borders as such, but that are needed – and set in motion or used – in the production chain to produce a certain good or service which is then eventually exported. For example, the land, energy and labour that is required to produce an agricultural good are not traded as such, but with the method used in this chapter, these outsourced 'upstream' or 'hidden' flows can be traced from the country of extraction/use back to the country of final demand.

5. Alf Hornborg. 'Ecological economics, Marxism, and technological progress: some explorations of the conceptual foundations of theories of ecologically unequal exchange'. *Ecological Economics* 105 (2014): 11–18, https://doi.org/10.1016/j.ecolecon.2014.05.015; Alf Hornborg. *Nature, Society, and Justice in the Anthropocene: Unraveling the Money-Energy-Technology Complex* (Cambridge: Cambridge University Press, 2019).

6. Frey R. Scott, Paul K. Gellert and Harry F. Dahms. *Ecologically Unequal Exchange: Environmental Injustice in Comparative and Historical Perspective* (Basingstoke: Palgrave Macmillan, 2018); Andrew K. Jorgenson and Brett Clark. 'Ecologically unequal exchange in comparative perspective: a brief introduction'. *International Journal of Comparative Sociology* 50 (3–4) (2009): 211–14.

7. Pablo Piñero, Martin Bruckner, Hanspeter Wieland, Eva Pongrácz and Stefan Giljum. 'The raw material basis of global value chains: allocating environmental responsibility based on value generation'. *Economic Systems Research* 31(2) (2019): 206–27; Christina Prell, Kuishuang Feng, Laixiang Sun, Martha Geores and Klaus Hubacek. 'The economic gains and environmental losses of US consumption: a world-systems and input-output approach'. *Social Forces* 93(1) (2014): 405–28, https://doi.org/10.1093/sf/sou048.

8. Gene M. Grossman and Elhanan Helpman. ' Quality ladders in the theory of growth'. *The Review of Economic Studies* 58(1) (1991): 43–61.

9. Christian Dorninger, Alf Hornborg, David J. Abson, Henrik von Wehrden, Anke Schaffartzik, Stefan Giljum, John-Oliver Engler, Robert L. Feller and Klaus Hubacek. 'Global Patterns of Ecologically Unequal Exchange: Implications for Sustainability in the 21st Century'. *Ecological Economics* 179 (2021): 106824, https://doi.org/10.1016/j.ecolecon.2020.106824.

10. Jason Hickel, Christian Dorninger, Hanspeter Wieland and Intan Suwandi. 'Imperialist appropriation in the world economy: drain from the Global South through unequal exchange, 1990-2015'. *Global Environmental Change* 73: 102467.

11. Manfred Lenzen, Arne Geschke, James West, Jacob Fry, Arunima Malik, Stefan Giljum, Llorenç Milà i Canals, Pablo Piñero, Stephan Lutter and Thomas Wiedmann. 'Implementing the material footprint to measure progress towards Sustainable Development Goals 8 and 12'. *Nature Sustainability* 5(2) (2022): 157–66.

12. Manfred Lenzen, Arne Geschke, Muhammad Daaniyall Abd Rahman, Yanyan Xiao, Jacob Fry, Rachel Reyes, Erik Dietzenbacher et al. 'The Global MRIO Lab – charting the world economy'. *Economic Systems Research* 29(2) (2017): 158–86, https://doi.org/10.1080/09535314.2017.1301887.

13. For more information on the method and data, see Dorninger et al. 'Global patterns'.

14. Christian Dorninger and Alf Hornborg. 'Can EEMRIO analyses establish the occurrence of ecologically unequal exchange?' *Ecological Economics* 119 (2015): 414–18, https://doi.org/10.1016/j.ecolecon.2015.08.009.

15. All monetary values are expressed in US dollars and deflated for the year 2015.

16. Samir Amin. 'Unequal development: an essay on the social formations of peripheral capitalism'. *Science and Society* 42(2) (1978).

17. Gernot Köhler. 'The structure of global money and world tables of unequal exchange. *Journal of World-Systems Research* 4(2) (1998): 145–68.

18. Where:

    T = value transfers through unequal exchange

    R_{net} = net resource drain from South to North

    P_N = Northern export price per resource unit

    M_{net} = net monetary transfers from North to South

19. Where:

    T = value transfers through unequal exchange

    R_{net} = net resource drain from South to North

    P_G = global average price

    M_{net} = net monetary transfers from North to South

20. Hickel et al. 'Imperialist appropriation', 5.

21. Alf Hornborg. *Global Magic: Technologies of Appropriation from Ancient Rome to Wall Street* (London: Palgrave Macmillan, 2016).

22. Hickel et al. 'Imperialist appropriation'.

23. Ibid., 9.

24. 'Global commodity chains and the new imperialism'. *Monthly Review* 70(10) (2019): 1–24.

7

Taking on the Eternal Debts of the South

Miriam Lang, Alberto Acosta and Esperanza Martínez

Resistance of the peoples who cry out for liberty!
It's time for reparations. This has to end.
Debts that bleed our land.
Floods, erosion cost the lives of the people... We owe nothing.
The rules are simple: debts or displacement...

Eternal debts enough![1]

Throughout history, debt has been one of the tools that has served to build and reaffirm all kinds of hierarchies and inequalities. Debt has been a powerful means of exploitation, subjugation and enslavement, which over time has taken on different disguises.

Conqueror indebtedness and indentured servitude were mainstays of the colonial system in the Americas: demanding high taxes, lending money with interest to those who couldn't pay, and requiring them to repay loans through work.[2] Along the same lines, recent research reveals that approximately a quarter of the people enslaved between 1600 and 1800 were enslaved as a result of indebtedness.[3] Later, after the abolition of slavery in the United States, racially discriminatory lending practices were systematically used to prevent the black population from equitably participating in society and the economy.[4]

Thus, the idea of 'race' served to 'scientifically' justify the differentiated exploitation of the workforce of black, indigenous or coloured people, expanding imperial power throughout the world and establishing different forms of control over bodies and subjectivities for each human group and gender. Debt also played a role in the subordination of workers and peasants; and in the separation between work considered to be 'productive' and work considered 'reproductive', a type of work that is normally unpaid, thus violently establishing spheres of the feminine and masculine. And debt was involved as well in the separation between human societies and the Nature in which they are inserted. In one way or another, most of these debts have appeared ever since the origins of modern states.

Debt comes into play to reinforce asymmetries between different actors and at different scales, reflecting patterns of domination and their intersections: class, race, gender, coloniality and society–Nature relations. As feminist readings teach us, despite the postulate of neutrality and abstraction that characterises the world of finance, debt exploits all social differences in a very concrete and situated way.[5]

There are multiple forms or types of debt: public, private, internal, external, commercial, monetary and non-monetary, personal, collective, family, colonial, racial, patriarchal, ecological or climatic. The concept itself is one of an obligation, whether moral or legal, to pay or return something. But throughout history, disputes have abounded, all revolving around a single question: who owes whom? And who has the power to define that?

This chapter aims to shed light on the effects and intersections of various forms of debt, prioritising the global frameworks and structural conditions that generate the interdependencies that lie at the base of the geopolitics of eco-social transition. Its main contribution to the debate raised by this book is that debt has given shape to the international division of labour that structures the relations between the Global Norths and Souths today, understood in their geopolitical, geoepistemic and geoeconomic dimensions and in their geographical heterogeneity. Thus, it must be understood as a powerful tool of domination causing structural global injustices.

THE WORLD ECONOMY: A CLEARLY TILTED PLAYING FIELD

The dominant narrative on debt, deeply rooted in our subjectivities, locates the responsibility for indebtedness as well as its consequences of all kinds in the popular classes, racialised social groups, or countries of the Global South, which are categorised as 'developing'. This is how the story goes: because of their inability, their irresponsibility, their deficiencies and their implicit inferiority, they have had to go into debt and are unable to pay back the commitments they acquired of their own free will; creditors generously 'help' them, which would make the debt legitimate. This narrative is built on the impossible assumption that the world economy is a level playing field, that success depends solely on the effort of each person – or each country

– and that, in the world market, the prices paid simply arise from the inter-action of supply and demand.

However, in practice, some debts end up becoming eternal. Latin American independence debt took a long time to be paid off, for some coun-tries, more than a century and a half. It transformed into a dependency debt, perhaps due to hereditary effects of an ancient curse that began more than 500 years ago, with the ransom of the Inca Atahualpa. In order to obtain his freedom, aware of the greed of the Spanish, the Inca monarch accepted the million-dollar ransom imposed by his captors, which, in the end, did not save him from being executed. Only thanks to the plundered gold and silver that arrived in Europe from Latin America between 1503 and 1660, a volume of 185 thousand kilos of gold and 16 million kilos of silver are recorded in the Archives of Seville. These resources, taken at today's value, would repre-sent an amount much higher than the total value of the external debt of all of Latin America.[6] It was only later, weighted down by the enormous debts of independence, that the countries of the region fully entered the vortex of external debt, with varying degrees of intensity and in the midst of a long and complex history of impositions and violence.

There are many illegitimate or even odious debts. Oftentimes, creditors knowingly take the risk of not being repaid on a loan. This was the case when, in the 1970s, the professed debt crisis came about, which ended up hitting Latin America so hard during the 1980s and 1990s. The high inter-national liquidity that gave rise to this crisis did not come about simply as a consequence of the increase in oil prices. Its real starting point was the economic impact of the Vietnam War and a trade dispute between the great powers. According to Aldo Ferrer, it was the United States that originated a new phase of the debt process: taking advantage of its dominant monetary position, it financed its economic imbalances by 'exporting' its national currency. Liquidity accumulated in the central banks and progressively filtered to private banks, increasing their capacity to grant credit.[7] Finan-cial resources grew through petrodollars, which, finding no productive use in the North, were channelled to the South, a place traditionally marginal-ised from international financial markets. The creditors did not take into consideration the debtors' ability to pay, since the financial business lies in lending, not in saving the dollars. Low interest rates that were lower than inflation constituted an invitation to continue borrowing. As David Graeber recounts, these loans 'started out at extremely low rates of interest that almost immediately skyrocketed to 20 percent or so due to tight U.S. money policies in the early '80s'.[8]

Alongside the banks, a multitude of foreign companies appeared on the scene, many of them transnational, which actively participated in the dance of the millions, even selling obsolete technologies or building works that due to design errors are true monuments to inefficiency. Examples of this include a US$2.5 billion thermonuclear plant in the Philippines on seismic ground which cannot operate; a paper mill in Santiago de Cao in Peru, which could not operate due to not having enough water; or a tin refinery in Karachipampa, Bolivia, which, because it is located at 4,000 meters above sea level, does not have enough oxygen to actually function.[9] These and many other projects today remain a liability to be paid for by poor countries. In many, the final cost was much higher than initially budgeted. Also, the purchase and sale of arms is often financed with external loans. The list of corruption cases related to foreign debt is enormous; they are a co-responsibility of the creditor and the debtor.

This is how those unpayable debts were created that later paved the way for the International Monetary Fund (IMF) to impose its structural adjustment programs, forcing the countries of the South to defund their health and public education systems, to abandon subsidies for basic foods, and to privatise their infrastructure – thus handing over much of their sovereignty and becoming dependent on new loans, getting caught up in a vicious circle. Meanwhile, the compound interest system guaranteed that, despite the fact that these countries had already paid back several times over the nominal amount they owed, their debt was not reduced.

As we have seen on multiple occasions in recent decades, while profits are privatised and concentrated in the hands of a few, the cost of servicing the debt is often distributed among the many. It becomes a tool for redistribution from the bottom to the top, a machine for widening inequalities. This was what happened in the case of the multiple bank bailouts, such as those of the 1990s in Mexico or Ecuador, or those induced in the US and Europe by the financial crisis of 2007/2008. It was during such crises when the irresponsible decisions of certain managers were compensated with huge sums of public money from taxpayers. That money could then no longer be used to satisfy the common needs of the people.

In short, the world economy is shaped by institutions that have highly unfair internal structures and by what, in practice, are exclusive clubs, namely the OECD, the G7 or the Paris Club. Imbalances are especially marked in global financial institutions such as the IMF and the World Bank (WB). In both of these bodies, the United States has *de facto* veto power over all major decisions and, along with the rest of the G7 and the European

Union, controls well over half the votes. While low- and middle-income countries are home to 85 per cent of the world's population, they have only a minority share of the votes when decisions are made at the WB or the IMF. Economic anthropologist Jason Hickel also points to a strong racial imbalance in the governance of world finance: on average, the votes of non-white people are worth only a fraction of the weight of those of their white counterparts.[10] Faced with this reality, it is not surprising that the Afro-American philosopher Olúfémi Táíwó characterises the capitalist world system as a 'global racial empire'.[11]

STRUCTURAL ADJUSTMENT AND POPULAR INDEBTEDNESS

In this context, with the imposition of neoliberal reason and an increasingly global economy, consolidation of a true debt economy has been underway, not only altering the forms of capital accumulation, but also class relations and the possibilities of political action. Talking about indebtedness in its contemporary form implies not only talking about public or sovereign indebtedness (the debt taken on by states), but also about the indebtedness of people in everyday life. Debt has become ubiquitous; today it affects billions of people very unevenly.[12] There is a broad consensus that the introduction of this debt economy was a strategy to counter the rise of social struggles in the 1960s and 1970s that threatened accumulation rates, a kind of counter-revolution of capital that imposed neoliberal reason on the economy and onto all other spheres of society.[13]

The dismantling of welfare state provisions in the North and the imposition of structural adjustment measures in the South led to a massive privatisation of the realm of reproduction. Through structural adjustment, the cost of servicing the foreign debt was transferred to the most impoverished sectors of the population, who in turn were pushed to take on debt to meet their daily needs, since education and even health services were only accessible through loans.[14]

This privatisation of the reproductive sphere coincided with the historical moment in which a significant proportion of women, previously confined to the 'private', invisible, and unrecognised world of reproductive and care work, finally gained access to paid work. Their inclusion in the paid labour market happened alongside the great neoliberal setback in labour rights and guarantees, throwing them into precarious conditions. Over time, the so-called 'non-bank' debts for medicine, rent, electricity, water, gas, etc., multiplied, while the proportion of women among the impoverished population grew. Silvia Federici, Verónica Gago and Lucy Cavallero confirm that there was also a kind of counterrevolution there, a response from above to feminist struggles: 'The invasion of finance in social reproduction that is directed especially at feminised economies responds to the feminist dispute for the recognition of historically devalued, poorly paid, and invisible tasks and a desire for economic autonomy.'[15] The so-called 'financial inclusion' or 'financial literacy' of the middle and popular classes not only put them at the mercy of banks and their conditions, but ended up depriving them of a common mechanism: between the nineteenth century and the second post-war period, as Federici points out, the worker culture included credit at the corner store or mutual loans to make it to payday, which was a way of circulating scarce resources in solidarity. This ended with the bancarisation of debt and the corresponding requirement of providing guarantees, which translates into a strategy of dispossession and expropriation.[16]

The fact that there is no other option than to go into debt to live, a reality further aggravated by the COVID-19 pandemic, operates as a productive tool, since it forces us to work more and more, while sucking up the energies for social struggle by individualising people through the feeling of guilt.[17] As a result, indebtedness has powerful gender implications. Not only is it specifically related to reproductive tasks, it also disproportionately affects women and feminised bodies, namely single mothers or heads of household, and is articulated with patriarchal violence.

CLAIMING THE ECOLOGICAL DEBT

At the end of the twentieth century, many voices arose from within Latin America to challenge the dominant discourse of debt and global injustice. It should be noted that some groups in the Global North also echoed their support for these claims, as happened in several European countries and in the United States itself. The structure of domination began to be questioned, even raising the claim of ecological debt.[18] 'We are not debtors, we are creditors!' was the motto of the Latin American and Caribbean Conference held in Ecuador in 2007, which stated that the external debt, already paid back multiple times over, is contrasted with an immense ecological, historical and colonial debt, which in reality the industrialised countries owe to the peoples of the South.[19]

In recent years, progress has been made in methodologies that make it possible to quantify, at least symbolically and for political purposes, this debt that the North owes to the South, in the context of what has been called unequal commercial and ecological exchange. Christian Dorninger's chapter in this book is an example. Other authors show, from a more historical perspective, that technological progress in the North in the aftermath of the industrial revolution was not the result of the genius of certain white men, but that some inventions rather were successful because they turned out to be very profitable at the time, thanks to the system of slavery that provided free labour and cheap raw material.[20] Other authors, in turn, disprove certain myths regarding North–South inequalities: for example, that exports from the North have a higher 'added value' because they incorporate higher labour productivity. In reality, the empirical data indicates that the real differences in productivity between workers are minimal and cannot explain the wage inequalities between the Norths and the Souths. Rather, these would be due to the greater, quasi-monopoly bargaining power of Northern companies and governments in determining the prices of their exports.[21]

All of the above leads to the conclusion that there will be no social justice without global justice and that proposing social justice at the level of individual countries is insufficient: nation-states never were and are not today closed circuits, but their prosperity or dysfunctionality have grown in dependence to one another throughout history. The prosperity of some, as dependency theorists already warned, was built on the plunder and subordination of others.

It is this interdependence that strongly determines the possibilities of an eco-social transformation that would be just in multiple dimensions, beyond the capacities or institutionality of a country or its population. In the words of Táíwò, 'a colonized nation is literally regulated by an external entity, and not by its internal system of government, while what a colonizing nation offers its citizens depends directly on what they extract from the colonies.'[22] A highly indebted nation finds itself in a very similar situation.

From the Conferences of the Parties (COPs) that have been held annually for three decades in the context of the 1992 United Nations Framework Convention on Climate Change and have failed to effectively reduce greenhouse gas emissions, a demand for climate justice has arisen. It is intended to show that the industrialised countries have for centuries borne a far greater responsibility for contributing to global warming than the countries of the South. In this context, the concept of climate debt was derived from the more comprehensive concept of ecological debt.

Although it was finally agreed at COP 27 in Egypt to set up a fund to compensate for the loss and damage of the countries most affected by climate change, it remains to be seen whether the implementation of this mechanism does not lead to new indebtedness of the recipient countries – for example, if the funds are provided as loans, which is a fairly common logic in what is called climate finance within the global financial system.

In a letter to the Transitional Committee that is working to propose how the Loss and Damage Fund should be operationalised and governed, climate justice groups from the Asian People's Movement on Debt and Development argue that: 'The recognition of climate debt and the Global South's right to reparations must be at the core of the Loss and Damage Fund. A huge climate debt is owed to developing countries by the Global North, corporations, and international financial institutions (IFIs) that have historically and currently been using more than their fair share of atmospheric space.'[23] According to Dorothy Guerrero from Global Justice Now, 'funds that will be released must come in the form of grants and not loans. Loss and Damage must not increase indebtedness. There is money, the countries and corporations that are most responsible for the climate crisis must be the ones to fill the fund. Fossil fuel conglomerates have hugely profited while creating so many climate impacts just in 2022 alone. This shows there is money to fill the fund.'[24]

In addition, debt-for-climate or debt-for-nature swaps have come back in vogue in many parts of the world. Those, as we know, neither solve the debt issue nor provide significant resources to support environmental policies. They are agreements that generally place the environmental policies of indebted countries under the control of international financial institutions and allow the most polluting countries to account for these swaps within their own 'Nationally Determined Contributions' under the 2015 Paris Agreement. This places all effective climate action on the shoulders of the debtor country, while the historical colonisers, creditors and polluters appear to have met their obligations. In the dispute over the mechanisms that will govern the loss and damage fund, it is key to include the perspective of ecological debt, as well as to consider the sovereignty of Southern countries over their own eco-social transition processes.

Although the notions of ecological debt and climate debt have circulated widely, and although many organisations and social movements have called for environmental justice and climate justice, no significant progress has been made in their political implementation. This would suppose, as the Eco-social and Intercultural Pact of the South recently proposed,[25] the annulment of the foreign debts of the South as a first step, based on the recognition of centuries of colonial appropriation and ecologically unequal exchange, both of which continue to this day.

Instead of acknowledging, in social and civilisational terms, the enormous contribution of Nature and feminised care work to the productive cycles and to sustaining life, the dominant economic thought does the opposite: in the name of 'valuing' these contributions, it has transformed them into new merchandise to extract monetary value and generate new opportunities for accumulation. Shortly after the Earth Summit in Rio 1992, the 'debt-for-conservation swap' mechanism was invented,[26] which links the supposed obligation of the countries of the South to pay their external debt with another additional obligation: the requirement to conserve their forests and biodiversity for foreign interests, for example in the context of emission offsets.[27] The external debt is nominally reduced. However, in exchange, groups of the environmental plutocracy of the debtor countries acquire, in complicity with the creditors, the right to interfere in its environmental policies, in the relationship of peoples with their territories, and also in the specific ways of performing conservation. As a result, this mechanism transforms into a new neocolonial appropriation strategy that reinforces power relations between creditors and debtors, disguised, as so often seen, as a 'win-win' solution. This results in the perpetuation of a perverse circle that begins with original accumulation, continues with accumulation by dispossession, and now adds on accumulation by conservation.

The prospects that the cancellation of external debts and the payment of the climate debt could be operationalised through the existing international institutions are also not very encouraging. It is true that at the beginning of 2020, under the initial impact of the pandemic, certain prominent figures that included Pope Francis and French president Emmanuel Macron proposed reducing or even cancelling the debt of the poorest countries.[28] That said, the geopolitical race between the major world powers encourages them to always and ultimately seek their own benefit.

At the subjective level, among the populations of the geopolitical North, there is a deeply rooted and implicit conviction that, somehow, they deserve to live better, with greater security against all kinds of threats, with better social benefits and more efficient institutions than the racialised populations of the South. This notion of implicit superiority, based on the introjection of the imperial mode of living, operates as a kind of cultural grammar, an invisible substratum that took shape over the centuries. The militant anti-immigration movements and parties are only the most radical expression of this feeling, which is based on what Aníbal Quijano called the colonial pattern of power. It is based on a cruelly tautological justification: because one lives better and it has always been so, 'naturally' one has the right to live better. This complicates the political viability of building global justice from the North.

QUESTIONING THE (IL)LOGIC OF 'ETERNAL' DEBTS: POSSIBLE REPARATION STRATEGIES

A first step towards global justice is, without a doubt, disobedience in the face of debt: getting organised so as not to pay, be it at the neighbourhood, national or regional level.

In order to make progress on collecting climate debt or environmental debt, both closely intertwined with the colonial debt, perhaps it is necessary to look at the issue from below, from a multiplicity of actors, instead of waiting for it to be resolved at the official multilateral level, which is increasingly dominated by corporations (see the chapter by Mary Ann Manahan in this book). Rethinking solidarities with the South requires leaving behind the 'logic of offsets', a logic which only 'works' when the Northern offsetters reap a benefit, such as, for example, when acquiring the right to continue polluting and emitting greenhouse gases (GHG).

The world to be built operates outside of possessive individualism, and its guiding principles are reciprocity and a sense of justice. The absolute reduction of GHG emissions is imperative and can only be achieved by leaving compensation dynamics behind. The order of the day should be to build horizontal relationships between towns, municipalities and organizations in the North and South that are willing to assume this perspective of ecological debt, starting with unconditional retribution for centuries of plunder. This in no way implies opening up to what is called 'corporate social responsibility'. Corporations, based on their legal framework, are entities strictly guided by profitability, and profitability is what has led us to collapse.

Another important aspect is to move away from the centrality of money that governs the capitalist civilisation. Although we talk about debt, be it colonial, environmental or climatic, the task is not just to organise large flows of money from North to South, or from country to country, or to communities or individuals. Money is the language of capitalism, and it serves in the first place to reinforce the capitalist relations that are exactly what we need to dismantle. Instead of restricting the gaze to the 'payment' of the debt, it is important to introduce the notions of restitution and reparation into the debate. Olúfémi Táíwó proposes a constructive perspective on reparations. Despite such reparations being motivated by past injustices, in their operationalisation, they do not aim at reconciliation or redemption. Rather, they aim to remake the world in other terms, with other rules of the game and other structures, to 'create a completely new political order, characterised by self-determination, non-domination and solidarity'.[29] This includes, of course, building a different type of economy[30] with radically different international economic structures.[31]

It is then a matter of restoring, insomuch as possible, what was removed or destroyed, in all its dimensions: material, environmental and symbolic. For example, restoring sovereignty to democratically make collective decisions about one's own future, in a situated and appropriate manner for each context in territories of cultural diversity; restoring sovereignty over the territory and food sovereignty; sovereignty over economic policy outside the yoke of foreign debt; restoring and recognising those modes of living that revolve around the quality of relationships and the balance between humans and Nature, instead of placing the accumulation of money and the concentration of power at the centre. In the same way that colonialism and the coloniality of power, the global economy and financialisation have generated a certain type of world,[32] we must build a resistance that is at the same time constitutive of another world.

NOTES

1. *Dívidas Eternas ¡Basta!*, produced by Acción Ecológica, video, 04:09, released 12 October 2020, www.youtube.com/watch?v=OnPQX6tv3cM.

2. David Graeber. *En deuda: una historia alternativa de la economía* (Barcelona: Editorial Ariel, 2014), p. 651.

3. Judith Spicksley. 'Debt, poverty and slavery', University of Hull, www.hull.ac.uk/work-with-us/research/institutes/wilberforce-institute/projects/debt-poverty-and-slavery.

4. Olúfémi Táíwò. *Reconsidering Reparations* (New York: Oxford University Press).

5. Lucy Cavallero and Verónica Gago. *Una lectura feminista de la deuda* (Buenos Aires: Rosa Luxemburg Foundation, 2019), p. 14.

6. Alberto Acosta. 'La increíble y triste historia de América Latina y su perversa deuda externa'. *La Insignia*, December 2002, www.lainsignia.org/2002/diciembre/econ_022.htm.

7. Aldo Ferrer. 'Deuda, soberanía y democracia en América Latina'. *Estudios Internacionales* 17(67) (1984): 309–23, https://doi.org/10.5354/0719-3769.1984.15797.

8. Graeber, *En Deuda*, 5.

9. Acosta, *Perversa deuda*.

10. Jason Hickel. 'Apartheid in the World Bank and IMF. Exploring banking, race and colonialism.' *Positive Money*, 7 January 2021, https://positivemoney.org/2021/01/apartheid-in-the-world-bank-and-the-imf/ (last accessed December 2022).

11. Táíwó, *Reconsidering*, 4.
12. Silvia Federici. 'From commoning to debt. Financiarization, Microcredit, and the Changing Architectures of Capital Accumulation.' *The South Atlantic Quarterly* 113(2) (2014): 231–44.
13. Edgardo Lander. 'La utopía del mercado total y el poder imperial.' *Revista Venezolana de Sociología* 8(2) (2002): 51–79.
14. Federici, 'From Commoning to Debt'.
15. Silvia Federici, Verónica Gago and Luci Cavallero (eds). ¿Quién le debe a quién? Ensayos transnacionales *de desobediencia financiera* (Buenos Aires: Tinta Limón and Rosa Luxemburg Foundation, 2021).
16. Federici, 'From Commoning to Debt', 234.
17. Federici et al., *Quien le debe*.
18. Alberto Acosta. 'La deuda externa acrecienta la deuda ecológica.' *Political Ecology Magazine* 14 (1997).
19. Ivonne Yánez and Aurora Donoso. *Sur, Soberanía Y Dignidad: No Somos Deudores, Somos Acreedores* (Quito: SPEDA-ALCA/Acción Ecológica/Instituto de Estudios Ecologistas del Tercer Mundo, 2008).
20. Alf Hornborg. *Global Magic. Technologies of Appropriation from Ancient Rome to Wall Street* (London: Palgrave Macmillan, 2016).
21. Jason Hickel, Christian Dorninger, Hanspeter Wieland and Intan Suwandi. 'Imperialist appropriation in the world economy: Drain from the global South through unequal exchange, 1990–2015.' *Global Environmental Change* 73 (2022): 1–13.
22. Táíwó, *Reconsidering*, 83.
23. Dorothy Guerrero. 'The UN Loss and Damage Transitional Committee's first meeting: what was discussed?' *Global Justice Now*, 30 March 2023, https://www.globaljustice.org.uk/blog/2023/03/the-un-loss-and-damage-transitional-committees-first-meeting-what-was-discussed-so-far/
24. Ibid.
25. Mónica Salvador et al. *Mecanismos de conversión de deuda – Alcances y limitaciones* (Quito: FONDAD CAAP, 1990); Martha Moncada and Juan Carlos Cuéllar. *El peso de la deuda externa ecuatoriana y el impacto de las alternativas de conversión para el Desarrollo* (Quito: Abya-Yala, 2004).
26. Ariel Salleh. 'Ecological debt: embodied debt', In *Eco-Sufficiency and Global Justice. Women Write Political Ecology* (London: Pluto Press, 2009).
27. Juliane Schumacher. *Green New Deals* (Brussels: Rosa Luxemburg Foundation, 2021).
28. Rodrigue Gozoa. 'Dette des pays pauvres: Macron échange avec le pape François.' *La Nouvelle Tribune*, April 2020, https://lanouvelletribune.info/2020/04/dette-des-pays-pauvres-macron-echange-avec-le-pape-francois/ (last accessed December 2022).
29. Táíwó, *Reconsidering*, 72.
30. Alberto Acosta and John Cajas Guijarro. 'Reflexiones sobre el sin-rumbo de la economía – De las "ciencias económicas" a la posteconomía.' *Ecuador Debate* 103 (2018); Alberto Acosta and John Cajas-Guijarro. 'Naturaleza, economía y subversión epistémica para la transición. Buscando fundamentos biocéntricos

para una post-economía, in Griselda Günter and Monika Meireles (eds), *Voces Latinoamericanas. Mercantilización de la naturaleza y resistencia social* (Mexico: UAM-Xochimilco, 2021).

31. Alberto Acosta and John Cajas Guijarro. 'Del coronavirus a la gran transformación – Repensando la institucionalidad de la económica global', in Pablo Amadeo (ed.), *Postnormales* (ASPO, 2020).

32. Adom Getachew. *Worldmaking after Empire: The Rise and Fall of Self-Determination* (Princeton, NJ: Princeton University Press, 2018).

8

What to Expect from the State in Social-Ecological Transformations?

Ulrich Brand and Miriam Lang

The easiest way to imagine just eco-social transitions is compiling a large list of public policies which would be implemented by national states. But how realistic is that, or how blind to historical experience and concrete societal power relations? Throughout history, emancipatory forces have met serious obstacles when betting on the state as the main actor of transformations. The historical experiences of anti-colonial movements and their 'becoming the state' are one prominent example: as they converted into state parties, anti-colonial forces in the Global South mostly gave up their former revolutionary and anti-capitalist ideals.[1] Another example are the limits which socialist politics have met in the Soviet Bloc, which, in retrospective, hardly can be framed as emancipatory. Social democratic parties and welfare states in the Global North would be a third example: they contributed to the material improvement of the living conditions of the masses and their political integration into a project of accelerated capitalist growth, at the expense of most of the Global South and of global ecosystems.[2] Particularly the experiences of progressive governments in Latin America between 1999 and 2014, but also of current leftist or centre-left governments there such as in Chile or Brazil, eloquently show that eco-social politics is not just a question of adequate state action.[3] What is hard to accept for emancipatory forces is that there is a clear tendency that, in moments of crises, the state intervenes rather in favour of the dominant economic and political interests.

Today, in those countries where 'green' capital and technologies are sufficiently developed and internationally competitive, we experience a transformation towards an 'eco-capitalist state'. This does not mean at all that the deepening ecological crisis and related injustices are resolved, but only that relevant fractions of capital, workers and their organisations, dominant science, the public, consumers and also the state, intend to deal with certain aspects of the ecological crisis. But this transformation tends to take place under conditions set by the dominant forces, a constellation that Gramsci called 'passive revolution'.[4]

On the other hand, in the Green New Deal proposals elaborated by different constellations of the Left of the Global North, political steering ideas are often shaped by (eco-)Keynesian perspectives. The state, or the EU institutions, are mostly understood as regulators, policymakers and as redistribution mechanisms that would advance socio-ecological transformation processes under corresponding left political and governmental conditions.[5] Current debates and policy strategies around a 'European Green Deal' or a – more leftist – 'Green New Deal' (GND) suggest – despite very significant differences – that the state might (re-)assume a far more prominent role by massively investing in infrastructure and renewable energy or by introducing emission-related border taxation.[6]

The fact that the state as a major interlocutor when it comes to social-ecological concerns constitutes a paradox that needs to be understood. Because, first of all, it needs to be acknowledged that the state is an important driver of the ongoing escalation of economic growth and unsustainable patterns of production and consumption.

The recent re-valuation of the state is mostly limited to the level of policy proposals – sometimes with a critical reference to the capital-friendly and repressive side of the state. But no critical understanding of the state (or supranational institutions like the EU) is developed. Its domineering, structurally selective and multi-scalar character is not assumed, and the political economy centred on the capitalist growth imperative is rarely questioned. This is the focus of this chapter, which also aims at developing arguments and concepts to better understand this vague and ambiguous entity that we call 'state'. To scrutinise these barriers to change, contradictions within dominant patterns that reproduce capitalist social relations and entry points for social-ecological transformations, is the strength of a critical thinking that avoids a state-centric view (which characterises most social scientific and philosophical approaches) and looks at the very societies which embed and enable the state, and are also, in parts, structured by it.[7]

UNDERSTANDING THE STATE

Of course, reflecting on 'the state' is difficult because the historically concrete manifestations of the state are very different.[8] The states of Bolivia, China, Germany, Russia or Tanzania are difficult to compare, even more from a critical perspective, where the state is not considered a more or less neutral regulator and framework-setter for the economy and society, but understood as interwoven with societies, cultures and economies. But despite those differences, which need to be acknowledged and will be discussed below, historical-materialist state theories help us to understand some central features of the capitalist state and its ambiguities, which we outline in the subsequent paragraphs.

To start with, the state is a relationship of domination, separated from, but intrinsically linked to, the capitalist economy and society. Its 'general function' – as Nicos Poulantzas put it – is to secure the conditions of the capitalist mode of production: the conditions for accumulation, such as the availability of wage-earners, natural resources, infrastructures, money and a certain order and stability.[9] At the same time, however, the state is not simply the 'instrument of capital' (because capital usually acts under conditions of competition with monopolistic tendencies) but maintains a certain 'relative autonomy' to secure capitalism as such – and not to serve distinct, concrete fractions of capital.

This becomes particularly clear when we think of the material basis which states have to rely on: it consists largely of taxes and tariffs collected from capitalist commodity production and the direct and indirect taxes of wage earners. Therefore, the state is no autonomous entity 'above' the rest of society, but an integral part of it. It also does not embody a pre-existing social 'general will', because social interests are far too divided. Many social struggles aim for certain interests to become 'generalised' through the state (for example, as part of the dynamics towards green capitalism and green colonialism), that means to be imposed on other interests as well, and to be promoted and secured through state policies.

Additionally, capital is not a homogeneous actor, but full of tensions and conflicts. Accumulation strategies might contradict each other, for example, be oriented more towards the world market or the internal market. Therefore, another central function of the state is to organise capital politically – and, in a certain way, to disorganise the subaltern classes. However, the interests of the wage-dependent people or the subaltern are also partially, and asymmetrically, inscribed into the state (as a result of previous struggles).

The state is dependent on a functioning capitalist economy, be it on the production of absolute or relative surplus, be it on rent, for example, the selling of concessions to mining or oil companies. The actually existing states are a part of the dominant capitalist growth regime, which enacts class-based, gendered and racialised, as well as global forms of domination and exploitation.

Historical-materialist approaches emphasise how the core structural principles of global capitalism permeate and shape state structures and processes: for example, private ownership over the means of production, implying competing and antagonistic social interests; or the inherent drive of capital towards accumulation through profit-maximisation due to competitive pressure. This corresponds with an inherent growth imperative, which is, in principle, indifferent to its ecological consequences. Under the conditions of neoliberal economic globalisation, transnational corporations compete on a global scale, and maintaining and increasing international competitiveness has become one of the primary policy goals of nation-states – often at the expense of stricter environmental regulation.

The state is not a neutral regulator, but highly interwoven with manifold social relations. Nor is it just one actor among others, as more recent 'governance'-debates suggest (see the chapter by Mary Ann Manahan on multistakeholder governance in this book). It is the focal terrain of societal power relations in which dominant social forces organise themselves by carrying out their conflicts in a rule-guided manner. As a consequence, power relations among social forces and specific political actors are historically inscribed into the political structure of the state – state institutions thus constitute a 'material condensation' of those power relations.[10]

This is also an entry point for emancipatory demands and forces. It has to be seen without any innocence. Emancipatory demands and related conflicts over particular politics or the general orientation of politics are fought out on a pre-structured, asymmetrical institutional terrain. This means that specific strategies and interests enjoy historically developed and entrenched privileged access to key decision-making areas; so-called structural and strategic selectivities.[11]

However, the state can not only be understood as an asymmetric terrain of struggles, but also as a system that can possibly block powerful interests and give emancipatory or eco-social demands and achievements certain durability:[12] leaving the oil in the soil, stopping the operation of nuclear power plants and the use of genetically modified organisms (GMOs), enabling the expansion of sustainable public transport and democratic energy transitions, creating rules and acting against gender or racialised violence, creating an education system that is part of the transformations we are talking about, introducing a tax system that supports them and so on. This can be promoted by the generation of binding rules, limiting destructive dynamics driven by existing power structures and dedicating resources to promote social-ecological processes such as the establishing of social-ecological provisioning systems and infrastructures that are not guided by profit.

Critical state theory also highlights an aspect that we might observe in everyday life, but which is often underestimated in analyses of the state: that various state apparatuses are in a tense relationship with each other, even contradicting each other in parts. It is therefore a political challenge to commit the various actors and organisations to one line, or to formulate and implement a reasonably coherent and realisable 'state project'.[13] The neoliberal movement has shown historically how such a state project can be formulated and implemented, and it is currently an open question whether green-capitalist forces are able to do this as well.

DIFFERENCES BETWEEN STATES IN THE GLOBAL NORTH AND SOUTH

As was already mentioned above, while a certain abstract form and logic of the state is assumed to be present in all historically concrete states, those, at the same time, differ considerably.[14]

The concrete form of the political economy in which it is embedded has consequences for the state. States on the peripheries of the modern capitalist world system play a role, for instance, in coordinating the exploitation of labour and channel the flow of natural resources to capitalist centres, in accepting or even imposing certain trade agreements, or in organising 'hard' currencies via credits and debts. Also, the countries of the Global South are characterised by a strong presence of international capital through foreign direct investment. The influence of foreign capital – as well as foreign political actors – is strongly inscribed into the state apparatuses, a phenomenon which Cardoso and Faletto called the 'internalisation of external interests'.[15] In sum, it is more difficult for (semi-)peripheral states to formulate ambitious transformative projects that are independent from foreign interests. Foreign capital only fuels processes of industrialisation under very particular conditions, as was historically the case in the biggest Latin American countries, Turkey, or China.

If the material basis of peripheral states comes from extractivist concessions and rents, the state bureaucracy and public servant jobs that create loyal followers are based on rent income and strengthen a tendency towards clientelism. In general, the state plays a greater role in the economy. In the peripheries, the peripheral/colonial state was structured on the basis of the super-exploitation of land and labour of racialised subjects to whom it denied citizenship until recently.[16] In this way, the peripheral state breaks even with the appearance of freedom of contract between formally equal owners, at least for a large part of its population. In the colonial South, the rule of law has historically served as a strategic tool to legitimise appropriation and plunder.[17] What was poured into laws was the exclusion of majorities along lines of ethnicity and later race, exclusion from paid work, the use of natural resources and political participation. This legacy undoubtedly complicates any political agenda toward eco-social transformations in the Global South even further. Peripheral forms of state are characterised by the precariousness of their institutional structures, arising from a series of internal and global power imbalances and associated relations of violence.[18]

On the other hand, another important difference between countries – and hence their states – of the Global North and the Global South is that in the latter, there are more spaces which are not totally permeated by capitalist logics, such as subsistence economies. Beside the capitalist mode of (re-)production, other modes remain important. Dependency theory called this simultaneous coexistence of capitalist and non-capitalist forms of (re-)production 'structural heterogeneity'. When it comes to political authority, those capitalist states also might coexist with other, often indigenous authorities. This is the principle of 'plurinational states', as, for example, Ecuador and Bolivia.[19] These forms of (re-)production can be understood as marginalised, but truly sustainable modes of living, which can and should inform strategies toward eco-social justice on the basis of their own epistemologies and practices (see the chapters of Tatiana Roa Avendaño and Pablo Bertinat as well as those of Arturo Escobar and Maria Campos in this book).

THE STATE AS A MULTI-SCALAR SOCIAL RELATION – THE INTERNATIONALISATION OF THE STATE

Our argument so far is that particularly the national state (and its ideological form as a nation-state) is a crucial instance to secure capitalist, patriarchal, racialised and international social relations, as well as societal Nature relations that are highly domination-shaped, unsustainable and unequal. And that the state is a highly asymmetrical terrain of contestation, which complicates eco-social transformation strategies centred on the national state and its public policies.

Yet, the national state plays a crucial role in the reproduction of social structures and processes and is itself such a structure and process, in other words: a social relation. Its enormous material resources, its legal, bureaucratic and coercive functions, its power to set norms and to sanction and the fact that the state is usually accepted by large parts of the people as state – despite all its dysfunctions or its repressive character– contribute to its outstanding role in our societies. The state has discursive power (e.g., framing constantly its decisions with 'progress', 'growth', 'migrants are a problem'), and it also generates knowledge about the society and the economy in order to steer and control it more adequately, that is, through demographic, economic or environmental data, studies, expert committees, etc. Moreover, the state is active within the economy and the provision of services, or the creation and maintenance of physical and social infrastructures, through state-owned or public enterprises; or as a state that receives rents out of the extraction of natural resources.

This is the reason why many political struggles are struggles over state and governmental power (which is not the same) and that political parties and struggles within and among them play an important role – in countries with single parties, such as China, of course, struggles within the party are decisive (and in times of dictatorship, parties usually do not play any role). This is also why the emancipatory left – being state-centred or state-critical, for good reasons – needs to engage with the state. Even emancipatory politics at a distance from the state have to deal with this really existing entity.

However, what needs to be considered is the multi-scalar character of the state. This is quite obvious and politically important when we look at sub-state levels, that is, regions and provinces, cities and administrative units of the countryside. There are also state structures and bureaucracies which have more or less decision-making and fiscal power. In the last years, many progressive and emancipatory experiences occurred at the municipal level or in (semi)rural regions.[20]

But we could also talk about a state at the international level, because particular rule-setting, conflict-dealing and bureaucratic modes have become increasingly internationalised, especially in what has been framed as global environmental governance. The same applies for certain state functions, particularly the securing of the conditions to reproduce capitalism politically, that is, via foreign investment rights or intellectual property rights.

The process of capitalist globalisation since the beginning of the 1990s led to what critical scholarship calls the 'internationalisation of the state'.[21] The creation of the World Trade Organisation (WTO) but also the upgrading of the World Bank and the IMF were an expression of this.[22] There has also been extensive international environmental legislation. In November 2020, the International Environmental Agreements Database Project (IEADP) listed 1,300 multilateral and 2,200 bilateral agreements.[23] So, by no means, the neoliberal globalisation meant a 'withdrawal of the state', it rather was enacted by the state.[24]

Globalisation and the internationalisation of the state was not something that simply occurred to nation-states, it was a process actively shaped by particularly dominant nation-states, establishing international state apparatuses, such as international organisations, regimes or governance networks. Based on the notion of the state as a condensation of a societal relation of forces, these international state apparatuses can be interpreted as a 'second-order condensation', that is, they are a condensation of force relations *between* nation-states (which are themselves a condensation of power relations at the national scale) as well as *between nation-states and other actors operating on a global scale* (see, again, the chapter by Mary Ann Manahan in this book).[25] A crucial benefit of this approach is that it allows to consider power asymmetries and competition between nation-states in global (environmental) governance, including the enduring colonial legacy of North–South asymmetries.[26]

EMANCIPATORY TRANSFORMATIONS AND THE ROLE OF THE STATE

The capitalist growth imperative – one of the main drivers of the multiple crisis – is not just an outcome of more and more production, consumption and capital valorisation. The growth imperative is also secured by and deeply inscribed into the state's structures.

Developing adequate political responses to the multiple crisis is not simply a question of overcoming abstract logics of path dependence, or of finding cooperative, techno-scientific solutions, but of confronting vested and highly organised interests in order to transform deeply rooted social relations of production, provision and consumption. A first step for leftist governments which aim at eco-social change would be creating a 'relative autonomy' from national oligarchic and transnational economic and political forces. But this implies the ability to change power relations, that is, to weaken oligarchies and the role of transnational capital in order to be able to formulate alternative societal and state projects, which might require alliances beyond the national level.

So, how to deal with the state in eco-social transformations? First, the very logics and structures of the capitalist, imperial, patriarchal, colonial and racist state needs to be entirely transformed; and this struggle also has to happen within the state.[27] Second, this will only be possible in conjuncture with social movements, conscious, organised and engaged people, critical debates from (social) sciences and progressive public and private businesses.

Third, the *principle of a state* – we could also call it *institutionalised public entities* (IPE) – at various scales, from the local to the global, is necessary.[28] This means that people establish structures and processes to deal with conflicts, to create binding rules and mechanisms of how to live together (including economically) in transparent and democratic ways, to acknowledge differences; some kind of institutions which give a certain durability to the manifold forms of living together, and are able to defend them when power and vested interests come in to act against the democratic principles that secure the good living for all, including Nature.

This is particularly the case if we also consider the need for some kind of democratic and transparent mechanisms of coordination at the global scale. Democratic structures and processes at local scales – as favoured by many bottom-up activists – are decisive, but not sufficient, because they have difficulties in blocking external powers. There is also a vast field of technological systems and existing infrastructures – communications, energy, water grids, etc. – which have to be dealt with democratically and in ecologically sustainable ways. Consciously engaging in the emancipatory shaping of the competences and powers of the different scales is a part of this struggle.

In this sense, the relations between state (public sphere, institutions, etc.), community and autonomy cannot be 'either/or'. These different dimensions are, and will continue to be, part of the world, and we have to deal with the tensions and complexities of their relationships.

We already highlighted that a crucial function of the state is to potentially block powerful interests and to give a relative durability to emancipatory demands and achievements.

Definitely, strategies which aim to go beyond green capitalism, green colonialism and the imperial speech (almost promise) of a 'just' transition largely limited to the Global North, need to rethink and remake the internal and international divisions of labour and Nature. A diversification and planned shrinking of economies, following strict eco-social criteria, is key (see also the chapters by Luis González Reyes and Bengi Akbulut in this book). This is not just about trade but about the very material forms and modes of (re-)production of societies, and it includes a remaking of material infrastructures, for example, for transport, energy provision and water, towards social and eco-social infrastructures that, in principle, enable a good living for all, not at the expense of others or of Nature.

What we can learn from past experiences and need to have in mind strategically: the very transformation of the state apparatuses themselves (the *polity*), its structures, processes, logics, selectivities and the personnel of bureaucracies at many levels is of utmost importance, at least as important as the transformation of *policies*. But this can only succeed if there is a change operated throughout society, a shift in the existing power relations which encompass the state. This cultural shift includes all sort of sociopolitical actors, as well as strong levels of organisation and mobilisation. The climate movements driven by younger generations since 2018 all over the world might, hopefully, be a first taste of this.

NOTES

1. Eric Hobsbawm. *The Age of Extremes: The Short Twentieth Century, 1914–1991* (London: Vintage Books, 1994).
2. Gurminder K. Bhambra and John Holmwood. 'Colonialism, postcolonialism and the liberal welfare state'. *New Political Economy* (2018), DOI: 10.1080/13563467.2017.1417369; Will Steffen et al. 'The trajectory of the Anthropocene: the great acceleration'. *The Anthropocene Review* 2(1) (2015): 81–98, https://doi.org/10.1177/2053019614564785.
3. Franck Gaudichaud et al. (eds). *Los progresismos latinoamericanos del siglo XXI. Ensayos de interpretación histórica* (México: UNAM, 2019).
4. Antonio Gramsci. *Selections from the Prison Notebooks* (New York: International Publishers, 1971); Magnus Ryner. *Passive Revolution/Silent Revolution: Europe's Recovery Plan, the Green Deal, and the German Question* (Helsinki: Global Political Economy Working Papers, 2021).
5. It is also surprising and politically highly problematic that almost all of those liberal and leftist approaches leave the international dimension aside and imagine eco-social transitions at the national or European Union scale.
6. European Commission. *The European Green Deal*. COM(2019) 640 final (Brussels: European Commission, 2019); See the overview of different Green New Deal proposals in Juliane Schumacher. *Green New Deals. A Big Deal for Fair Climate Protection or Just the Latest Version of the Capitalist Model?* (Brussels: Rosa Luxemburg Foundation, 2021).
7. See also Miriam Lang. 'El rol del Estado en la transición hacia una sociedad post-extractivista: aportes para un debate necesario'. *Ecuador Debate* 117 (2022): 143–69; Ulrich Brand. 'How to get out of the multiple crisis? Towards a critical theory of social-ecological transformation'. *Environmental Values* 25(5) (2016): 503–25.

8. Moreover, in the state-centric debates the state is often confused with the government assuming a mere top-down process of policymaking and letting the whole complexities, conflicts and contingencies of the state apparatuses aside.

9. Nicos Poulantzas. *State, Power, Socialism* (London: Verso, 1978).

10. Ibid.

11. Bob Jessop. *State Power: A Strategic-relational Approach* (Cambridge: Polity Press, 2007).

12. Miriam Lang and Ulrich Brand. 'Dimensiones de la transformación social y el rol de las instituciones', in Miriam Lang, Belén Cevallos and Claudia López (eds), *Cómo transformar? Instituciones y cambio social en América Latina y Europa* (Quito: Abya Yala/Fundación Rosa Luxemburg), pp. 7–32.

13. Jessop. *State Power*.

14. Lang, *El rol.*; Álvaro García Linera, Raúl Prada, Luis Tapia and Oscar Vega Camacho (eds). *El Estado. Campo de lucha* (La Paz: Clacso et al., 2010); Tobias Boos and Ulrich Brand. 'State transformation', in Olaf Kaltmeier, Anne Tittor, Daniel Hawkins and Eleonora Rohland (eds), *The Routledge Handbook to the Political Economy and Governance of the Americas* (London: Routledge, 2022), pp. 221–30.

15. Fernando H. Cardoso and Enzo Faletto. *Dependency and Development in Latin America (Dependencia Y Desarrollo En América Latina, Engl.)* (Berkeley, CA: University of California Press, 1979).

16. Aníbal Quijano. 'Coloniality of power, eurocentrism, and Latin America.' *Nepentia. Views from the South* 1(3) (2000): 533–80.

17. Ugo Mattei and Laura Nader. *Plunder. When the Rule of Law is Illegal* (Malden, MA, Oxford and Carlton: Blackwell Publishing, 2008).

18. Lang, *El rol.*

19. Miriam Lang. 'Plurinationality as a strategy. Transforming local state institutions toward Buen Vivir', in Elise Klein and Carlos Eduardo Morreo (eds), *Postdevelopment in Practice. Alternatives, Economies, Ontologies* (New York and London: Routledge, 2019), pp. 176–89.

20. See, for example, Mabrouka M'barek, Giorgos Velegrakis et al. (eds), *Cities of Dignity. Urban Transformations around the World* (Brussels: Rosa Luxemburg Foundation, 2019).

21. Ulrich Brand, Christoph Görg and Markus Wissen. 'Second-order condensations of societal power relations. Environmental politics and the internationalization of the state from a neo-Poulantzian perspective.' *Antipode* 43(1) (2011): 149–75; Birgit Sauer and Stefanie Wöhl. 'Feminist perspectives on the internationalization of the state.' *Antipode* 43(1) (2011): 108–28. 'Internationalisation of the state' does not only refer to the increasing importance of international state apparatuses but also of a more international orientation of the national apparatuses such as the strict orientation at 'international competitiveness'.

22. Quinn Slobodian. *Globalists: The End of Empire and the Birth of Neoliberalism* (Cambridge, MA: Harvard University Press, 2018).

23. See International Environmental Agreements Database Project (IEADP), https://iea.uoregon.edu/.

24. Joachim Hirsch. 'Globalization of capital, nation-states and democracy.' *Studies in Political Economy* 54(1) (2016): 39–58.
25. Brand et al. *Second-order condensations.*
26. Chukwumerjie Okereke, Doris Fuchs and Anders Hayden. 'North–South inequity and global environmental governance', in Agni Kalfagianni et al. (eds), *Routledge Handbook of Global Sustainability Governance* (London: Routledge, 2021), pp. 167–79.
27. Lang, *El rol.*
28. Lang and Brand, *Dimensiones.*

9

Green Colonialism in Colonial Structures: A Pan-African Perspective

Nnimmo Bassey

When Africa is seen as a state, ignoring the fragmentation of the continent into many countries, the point of the inherent multiplicity of realities and subjectivities is missed. Africa may have been a country only at the origin of humans and not much longer after that. As territories are more than geographic expressions, we must keep in mind the multiple realities of Africa. The struggle to forge a Pan-African reality has illustrated the multilocational spread of Africa, including through the understanding of who Africans are. It helps to see the configuration of Africa both in terms of geography and widely beyond physical boundaries. This framing helps for an understanding of the huge influences of the diaspora on the struggles for true liberation of the continent and points at an inclusive reconstruction that on the principles of gender and ecological justice, with clear eco-socialist underpinnings.

We do not have to paint an idyllic picture of life in Africa. Still, the continent had a history of kingdoms and advanced cultures before the disruptive arrival of European adventurers and traders in the fifteenth century. While we do not wish to reconstruct those moments, it is pertinent to note that the succeeding centuries birthed violent clashes, became entrenched through colonialism and persist to this day. The visitors met largely orderly communities and kingdoms with social arrangements built on respect, interdependency and care.

Life in the communities was ruled by a state of embeddedness – an understanding that human agency is governed by our interconnections with the world around us. According to Omedi Ochieng, this embeddedness involves acknowledging the ecological and historical background that structures and forms what we are and shapes the horizons of what we could become. It begins with an attentiveness to the air we breathe, the land we walk upon, the water we drink, the fire we use, and the formidable weight of history – politics, economics and culture.[1] These have been largely crushed or distorted by years of conquest, colonialism and imperial expropriation.

The view of Africa as a vast landscape with limitless resources to be exploited got rigidly entrenched in the colonial mindset and drove concepts of conservation that ignored the true cause of the degradation that unfolded from their mindless exploitation and consumption. Such concepts have been understood by some as fortress conservation,[2] requiring the removal of people from African landscapes ostensibly to 'protect nature'. This invention of the colonial mindset has given rise to grave social injustices in attempts to recreate mythical 'African Edens'.[3]

To the colonialist, the colonised territories should be areas devoid of humans and human communities. This posture is rooted in the vision of Africa or any colonised territory as a sacrificial, empty territory which is applicable only in the terms set by the colonialist. As stated in my 2012 book *To Cook a Continent*, the question often peddled in policy circles is: what can be done about Africa? And, in moments of generosity, the question moves to: what can be done for Africa?[4]

GREEN AND INTERNAL COLONIALISM IN AFRICA

In the context of green colonialism, we must also look at: what has been done to Africa? Green colonialism is a merger and extension of political, economic and socio-cultural colonialism. It has been built and cemented on the deep-seated coloniality through which African leaders have been programmed to believe, for example, in the international system of heritage conservation, and have utilised so-called international or alien standards to promote their own interests.[5]

Besides fortress conservation, colonialism sold the local elites the idea of looking to external economies for cash in exchange for natural materials and labour. Neo-colonial states continue this pattern of seeking foreign direct investments (FDIs) which primarily extract labour and raw materials and give them foreign exchange whose values are remotely set. Examples of how colonies got trapped in these foreign exchange dead ends can be seen in plantation agriculture, which shifted cropping for food to cropping for cash. Cash crop agriculture in the colonial era continued exploitative agricultural systems built under slavery. Today, plantation agriculture continues to produce export crops, triggering land grabs and excluding the farmers from producing food for their communities. To complicate the matter, plantation and monocropping, besides feeding external markets, now also provide biofuels for machines or bioenergy. Whether in the agricultural, mining or fossil fuels sectors, African leaders continue to look out mainly for foreign exchange, for prices they play no role in setting.

Structures erected by colonialism and the postcolonial era dramatically altered the socio-economic and political dynamics of the African continent. The seeds for rent-seeking patterns were sown by colonialism and watered by the manipulations of international financial institutions such as the World Bank and the International Monetary Fund (IMF). Debt has also been a tool for altering developmental imaginaries and pressuring countries to open more to plunder (see the chapter of Miriam Lang, Alberto Acosta and Esperanza Martínez in this book). Governments are under pressure to service external debts and meet import requirements and give transnational corporations liberal economic conditions, including tax breaks, labour quotas and the freedom to repatriate all profits in their transactions. They also engage in incestuous partnerships with these corporations, making it impossible to institute serious regulatory oversight. The governments' unwillingness and/ or inability to control corporations' actions have led to ecocidal exploitation, which has already created dead zones in some areas.

The consolidation of the freedom to exploit has also been aided by the creation of free trade or special economic zones, which have been characterised as enclaves of exception (see Rachmi Hertanti's chapter in this book).[6] One class of free trade zone (FTZ) is the export-processing zone (EPZ), usually set up in developing countries by their governments to promote industrial and commercial exports. According to the World Bank, these zones are 'small, fenced-in, duty-free areas, offering warehousing, storage, and distribution facilities for trade, transhipment, and re-export operations.[7] Many countries see those zones as the primary stimuli for attracting foreign direct investments. The United Nations Conference on Trade and Development (UNCTAD) reports that over 200 special economic zones (SEZ) are spread across 38 African countries. It also notes that 'at least 56 zones are under construction, and others are still at an early stage of development.[8] About 150,000 hectares of land in Africa are dedicated to SEZs, while over US$2.6 billion have been mobilised in investments into agro-processing, manufacturing and services.[9]

The regime of extraction for foreign exchange has been one never ending story of subtraction, adding scant value to the people or planet. Perfunctory voluntary human rights principles and transparency initiatives help corporations to greenwash their activities and export dirt on 'corrupt politicians'.

This unfortunate situation was foreseen by Frantz Fanon when he noted in his classic book *The Wretched of the Earth* that colonialism contents itself with bringing to light the natural resources, which it extracts and exports to meet the needs of the mother country's industries, thereby allowing certain sectors of the colony to become relatively rich. 'But the rest of the colony follows its path of under-development and poverty, or at all events, sinks into it more deeply.'[10] Fanon saw how colonial structures fragment nations and widen subjectivities that put the brakes on efforts to build African unity. He noted that:

African unity, that vague formula, yet one to which the men and women of Africa were passionately attached, and whose operative value served to bring immense pressure to bear on colonialism, African unity takes off the mask, and crumbles into regionalism inside the hollow shell of nationality itself. The national bourgeoisie, since it is strung up to defend its immediate interests and sees no farther than the end of its nose, reveals itself incapable of simply bringing national unity into being or building up the nation on a stable and productive basis. The national front which has forced colonialism to withdraw cracks up and wastes the victory it has gained.[11]

Our reading of Fanon clarifies how the political elite gets to see themselves as producers of niches of opportunity for their nations and rent-seeking as the engine for progress. This explains why current leaders are so stuck with the position that exploitation of fossil fuels and other minerals for export/ cash is a right that cannot be negotiated. This also locks in the specious notion that 'ecocide must be accepted as omelettes cannot be made without breaking eggs'.

Rapacious exploitation requires a thorough rethinking of development. The role of the World Bank and the IMF in enforcing the defunding of social services including health, education as well as economic supports, through their infamous structural adjustment programmes, stand out as colonial manipulations that upended common sense, reversed progress, instituted poverty and constructed underdevelopment. The perverse influence of these institutions underscores the need to pay close attention to the inequalities in power, using an eco-socialist and anticolonial lens.

Lessons on the political dynamics arising from colonial and postcolonial structures can be learned from Mozambique, one of the last countries to gain independence in Africa. The nation became independent in 1975.

This was followed by a cascade of phases, including dismantling most of the colonial structures and setting up a communist/socialist system. The nation experienced 16 years of civil war and had its first election in 1994. That election was supposed to bring in a democratic era. But, according to Anabela Lemos,

> this new democracy was nothing more than the complete opening of the doors to free markets and to all types of foreign investments in the name of development. This 'investment' started to come in different forms, such as massive aluminium smelters, large scale monoculture plantations of exotic trees, massive introduction of cash and export crops, and many others. A 'mining' economy was introduced, where everything was taken out of the country.[12]

SCRAMBLE FOR FOSSILS

In the case of the scramble for African oil and gas, the leaders see only an opportunity for their countries to benefit from fast-tracked projects. The argument is that expanded production would boost energy access for their people, even though this is a fatuous claim, given that decades of extraction have yielded only ecological devastation and poverty.[13]

The fixation on colonial trade had built what may also be termed voodoo economics. In this system, cash flows in with little production or transformation of raw materials. This has entrenched a culture of rentism or dependency whereby African countries depend on multinational extractive corporations for their national revenue. It is no surprise that oil revenues represent at least 20 per cent of GDP in Libya, Algeria, Gabon, Chad, Angola and The Republic of Congo. And although oil and gas contribute a modest 6 per cent of real GDP of Nigeria, they account for 95 per cent of foreign exchange income and 80 per cent of government revenues.[14] The African Union group of nations used the COP27 climate negotiations in Sharm-el-Sheikh 2022 to lobby for the expansion[15] of fossil fuel production to benefit from vast resources, as richer nations have supposedly done. Their argument is bereft of critical examination of the appropriation and externalisation mechanisms that made it possible for the richer nations to benefit from vast resources.

The root of the resource grab in Africa cannot be extricated from colonialism as it provided the base for impunity without fear of being held to account. Plunder and impunity have grown over the years, with brutal force when necessary. This has made the map of natural resources and conflicts on the continent overlap almost perfectly. Exploitation has been backed by national armies, special security agents and mercenaries. Extraction is literally carried out behind military shields, ignoring human and collective rights.

Patrick Bond, a political ecologist, aptly captures the troubling situation of the endless push for fossil fuels in the face of global warming using the role of France, South Africa and Rwanda. 'Total's current operations in Africa follow an old pattern: fossil fuel exploitation and corruption of developing country economies, governments, societies and environments, all backed by French state power.' As support for his assertion, he stated that: 'Emmanuel Macron [the president of France] made this abundantly clear in 2021 when he insisted on defending Total's US$20 billion gas assets in Mozambique through military intervention, led by Rwandan and South African soldiers. Pretoria's sub-imperialist role explains its desperate support for the new oil tycoons with whom Total has been allied since the mid-2010s to exploit large gas reserves and search for new deposits by seismic blasting.'[16] Bond notes that two forms of resistance have emerged against the revival of fossil imperialism and sub-imperialism in this axis since 2021: violent conflict that has shaken Total, the French oil and gas giant; and environmental and social mobilisations on the South African coastline that have rattled that country's government.

The role of France, a country that maintains a strict colonial grip on Francophone nations in Africa, is especially interesting. While France has outlawed fracking and crude oil extraction on its territories[17] and also has banned fossil fuels advertisements,[18] its oil and gas behemoth, TotalEnergies, continues to extract elsewhere and most notoriously at Cabo Delgado, Mozambique from where the first shipment[19] of fossil gas took place as COP27 was happening in Sharm el-Sheikh. The timing of the first shipment illustrates how violence has not stopped resource extraction in Africa as they often go in concert. This is epitomised in cases of the blood diamonds of Liberia as well as the ongoing instability in the Democratic Republic of Congo.

Total is one of the biggest players in the gas extraction at Cabo Delgado. The onshore Afungi LNG Park, built for the fossil fuel business, has led to the displacement of over 550 families in order to build a 70 kilometres road to the Park which has an aerodrome as well as treatment plants and port facilities. Coastal fishing communities have been displaced to a 'relocation village' that is more than 10km inland, effectively cutting them off from the sea and denying them their farmlands, fishing grounds, general livelihoods, culture and everything that matters to coastal communities.[20] Cabo Delgado hosts Africa's three largest liquid natural gas (LNG) projects: the Mozambique LNG Project (Total, formerly Anadarko) with a value of US$20 billion, Coral FLNG Project (ENI and ExxonMobil) with a value of US$4.7 billion, and Rovuma LNG Project (ExxonMobil, ENI and CNPC) valued at US$30 billion.[21] Cabo Delgado may be the site of one of the continent's biggest corporate-made disasters.

At a meeting on corporate impunity[22] hosted by Justiça Ambiental in Maputo, a community person declared very poignantly: 'For us, the multinational corporations did not bring development, they brought disgrace.' Substitute 'multinational corporations' with 'colonialism' and a fuller picture emerges. Another delegate at the meeting wondered if the destruction of their land could be called development. He then asked rhetorically: 'Is that the development we want?'

Colonialism, whether black, blue or green, never consults with the people. This lack of consultation is bred by an ingrained lack of respect for the people and planet. Playing the colonial game, areas where Total, the oil and gas company, operates, are suffering from a rise in social inequalities and resultant divisions, with the only unifying factor being that they are generally known as Total Areas.

GAS GRABBING IN AFRICA AND THE UPENDING OF THE CLIMATE CRISIS

The war in Ukraine and a general disregard for the need to take real climate action have led to massive recent investments in the fossil fuels sector in Africa. The current scramble for new oil, gas and coal projects by oil companies is ongoing in 48 African countries. When it has become clear that all of the fossil fuels from known reserves cannot be extracted and burnt without breaching the climate tipping point, oil companies plan to sink US$230 billion on new oil and gas projects in the next decade and US$1.4 trillion by 2050.[23] As if to cheer on the fossil corporations, an African entrepreneur, NJ Ayuk insists that 'Boycotting oil and gas firms in Africa is a misguided course of action.' Gabriel Obiang Lima, Equatorial Guinea's minister of mines and hydrocarbons, swore that: 'Under no circumstances are we going to be apologizing.... Anybody out of the continent saying we should not develop those [oil and gas] fields, that is a criminal...'[24] Despite the swashbuckling insistence on fossil fuels cash, the fact remains that 17 of the 20 countries most vulnerable to climate change are located in Africa, and the continent requires 'funding to help it adapt to the economic and humanitarian challenge of repeated climate disasters'.[25]

As though waiting for the whistles to be blown, the fossil fuels industry is doing everything possible to root the impression in the collective imagination that dependence on fossil fuels, even in the future, is inevitable. African leaders are fully aligned with this imagination and join fossil fuel speculators elsewhere to underline the fact that the war against Ukraine is an opportunity to lock in fossil fuel dependency. This is a convenient argument because the world cringes at the horrors of the exposure of Ukrainians to a cold winter. In addition, both parties in the conflict also need more fuels to run their war machinery.

Fossil fuel corporations can take the conflict in Cabo Delgado, Mozambique, in their stride. Building a 1,400km heavy crude pipeline from Uganda to export facilities in Tanzania is only a game. They can propose pipelines of conflict from the Niger Delta, Nigeria, to Morocco and another across the Sahara Desert to Algeria to feed the energy needs of Europe. With 89 per cent of gas infrastructure in Africa being for export purposes, meeting the energy needs of the African continent itself is missing from the cards.

DECOLONISING THE TRANSITION

A transition to green energy neither assures justice nor breaks away from the colonial pattern associated with dirty energy. This pattern is already emerging with the massive solar farms in Morocco[26] and the wind farm installed at Lake Turkana[27] in northern Kenya. The issues of concern here include the uptake of land, displacement of communities from the gifts of Nature they had enjoyed for millennia and the energy poverty that the installations do not eliminate, due to access challenges and because they are designed for export and not for the immediate host communities. Thus, the colonial pattern of land grabbing and lack of consultation can persist all the same under green energy scenarios (see also Hamza Hamouchene's chapter in this book).

Another sticking point is related to the extraction of the minerals needed to construct parts of the equipment required for green or renewable energy systems. The mining of the materials, including for the batteries required for energy storage, continues to impact communities and territories. The implication of this is that transition is not merely about energy, but about the need to decolonise the entire energy, economic and political systems.

GOING FORWARD

Although the lines separating the formal and informal sectors are sometimes blurred, informal or unofficial formations hold the key to decolonising and liberating the African continent and building an eco-socialist future. The informal sector must embark on severe engagements and confrontations with the system assimilated by colonial powers whose imaginaries have been disconnected from independent possibilities and a future with dignity and respect. This will always be a challenging task, as vested interests are keen to maintain the exploitative grip on the levers of power. Consider what happened in Guinea when then President Sékou Touré held a referendum in 1958 on whether to join the proposed Communauté Financière Africaine (CFA) monetary union. The votes came in on a resounding 95 per cent rejection of the union. Then, according to David Hundeyin, 'Charles de Gaulle's government immediately pulled out more than 4,000 civil servants, judges, teachers, doctors, and technicians, instructing them to sabotage everything they left behind.'[28] There have been fewer blatant acts of sabotage and brigandage over the years, but the impacts have all added to impede progress on the continent.

African governments are seen to increasingly work in cahoots with transnational corporations and philanthro-capitalists at multilateral fora such as the Conference of Parties to both the United Nations Conventions on climate change and biodiversity. Unlike what was the case a few decades ago, today these governments adopt market-based false solutions in a bid to attract financial support. It appears that gone are the days when the African Union (then Organisation of African Unity – OAU) had a strict model law to aid African countries to ensure biosafety in their countries. It appears that gone are the days when an African negotiator broke down in tears at a press conference at the climate COP15 in Copenhagen[29] because they were being pressured to sign a 'suicide pact'.

But not so soon. On the continent, there is a strong stirring of mass movements on climate change, food sovereignty and environmental justice. There are strong voices from below standing against ecocide and defending their forests and ocean. The Ogoni people of the Niger Delta stand as an example of a people that have resolutely expelled oil majors from their lands and swamps, rejecting the opening of oil wells[30] in their territory and forced a clean-up programme of hydrocarbon pollution in their environment after an assessment by UNEP.[31] We see communities along the coastline of South Africa standing firm against seismic exploration for crude oil and gas, defending their livelihoods, culture and spirituality. We see youths standing up against pipelines, fishers and forest-dependent communities defending their rights. The story is the same with farmers stoutly defending food sovereignty and rejecting seed colonialism, defending their farmlands against oil pollution.

The horizon brims with both struggles and hope. A coalescing of forces with clear political analyses are needed to build a truly united Africa with internal economic cohesion and full respect for the rights of the peoples and of Mother Earth.

NOTES

1. Omedi Ochieng. 'What African philosophy can teach you about the good life', iai.tv, 10 September 2018, https://iai.tv/articles/what-african-philosophy-can-teach-you-about-the-good-life-auid-1147 (last accessed December 2022).

2. Hannah Marcus. 'Fortress conservation: the green colonialism that must end to achieve ecological harmony in a post-COVID world.' *Planetary Health Alliance*, 13 March 2021, https://phalliance.medium.com/fortress-conservation-the-green-colonialism-that-must-end-to-achieve-ecological-harmony-in-a-41010b63 1c6f (last accessed December 2022).

3. Guillaume Blanc. 'The invention of green colonialism – the roots of Africa's wildlife NGOs come under withering scrutiny.' *Land Portal*, 14 June 2022, https://landportal.org/node/102593.

4. Nnimmo Bassey. *To Cook a Continent – Destructive Extraction and the Climate Crisis in Africa* (Oxford: Fahamu Press, 2012), p. viii.

5. Guillaume Blanc. 'The invention of green colonialism – Putting an end to the myth of the African Eden', interview by Flora Trouilloud, iD4D, 7 January 2021, https://ideas4development.org/en/green-colonialism-western-outlook/ (last accessed December 2022).

6. Omolade Adunbi. *Enclaves of Exception – Special Economic Zones and Extractive Practices in Nigeria* (Bloomington, IN: Indiana University Press, 2022).

7. World Bank. *Special Economic Zone: Performance, Lessons Learned, and Implication for Zone Development* (Washington, DC: World Bank, 2008), pp. 9–11.

8. UNCTAD (2021).

9. Busola Aro. 'AfCFTA: Africa must compete favourably with other free zones, says Buhari.' *The Cable*, 2 December 2022, www.thecable.ng/afcfta-africa-must-compete-favourably-with-other-free-zones-says-buhari (last accessed December 2022).

10. Frantz Fanon. *The Wretched of the Earth* (New York: Grove Press, 2004), p. 106.

11. Ibid., p. 110.

12. Anabela Lemos. 'Africa: the oil new frontier'. *FightTheFire*, 2022, www.fightthefire.net/africa-the-old-new-frontier/ (last accessed May 2023).

13. Nnimmo Bassey. 'The colonial exploitation of Africa's fossil fuels must stop.' *Context*, November 2022, www.context.news/net-zero/opinion/the-colonial-exploitation-of-africas-fossil-fuels-must-stop (last accessed December 2022).

14. NJ Ayuk. 'Africa must oppose measures at COP27 that restricts its fossil fuels.' *BusinessDay*, 7 November 2022, https://businessday.ng/news/article/africa-must-oppose-measures-at-cop-27-that-restricts-its-fossil-fuels/ (last accessed December 2022).

15. Fiona Harvey. 'African nations expected to make case for big rise in fossil fuel output.' *The Guardian*, 1 August 2022, www.theguardian.com/world/2022/aug/01/african-nations-set-to-make-the-case-for-big-rise-in-fossil-fuel-output (last accessed December 2022).

16. Patrick Bond. 'French fossil imperialism, South African subimperialism and anti-imperial resistance.' *CADTM*, December 2022, www.cadtm.org/French-fossil-imperialism-South-African-subimperialism-and-anti-imperial.

17. Agence France-Presse. 'France bans fracking and oil extraction on all of its territories', *The Guardian*, 20 December 2017, www.theguardian.com/environment/2017/dec/20/france-bans-fracking-and-oil-extraction-in-all-of-its-territories

18. Rosie Frost. 'France becomes the first European country to ban fossil fuel ads – but does the new law go far enough?' *Euronews*, 24 August 2022, www.euronews.com/green/2022/08/24/france-becomes-first-european-country-to-ban-fossil-fuel-ads-but-does-the-new-law-go-far-e (last accessed December 2022).

19. 'Mozambican leader announces first LNG export shipment.' *Africanews*, November 2022, www.africanews.com/2022/11/13/mozambican-leader-announces-first-lng-export-shipment// (last accessed December 2022).

20. Ilham Rawoot and Daniel Ribeiro. 'A total mess.' *Stopmozgas*, 27 July 2022, https://stopmozgas.org/article/total-mess/ (last accessed December 2022).

21. Ilham Rawoot. 'Gas-rich Mozambique may be headed for a disaster.' *Aljazeera*, 24 February 2020, www.aljazeera.com/opinions/2020/2/24/gas-rich-mozambique-may-be-headed-for-a-disaster/ (last accessed December 2022).

22. This meeting held over one week in November 2022 was attended by over 100 community persons from all over Mozambique.

23. NJ Ayuk. 'Africa will develop with oil and gas – whether the West likes it or not.' *World Economic Forum*, 19 January 2020, www.weforum.org/agenda/2020/01/africa-oil-gas-development/ (last accessed December 2022).

24. Libby George and Shadia Nasralla. 'No apologies: Africans say their need for oil cash outweighs climate concerns.' *Reuters*, 8 November 2019, www.reuters.com/article/us-africa-oil-climate-idUSKBN1XI16X (last accessed December 2022).

25. Oliver Milman. 'Two-thirds of US money for fossil fuel pours into Africa despite climate goals.' *The Guardian*, 31 October 2022, www.theguardian.com/environment/2022/oct/31/two-thirds-of-us-money-for-fossil-fuel-pours-into-africa-despite-climate-goals (last accessed December 2022).

26. Joanna Allan. 'Renewable energy is fuelling a forgotten conflict in Africa's last colony.' *The Conversation*, 26 November 2021, https://theconversation.com/renewable-energy-is-fuelling-a-forgotten-conflict-in-africas-last-colony-170995 (last accessed December 2022).

27. Business & Human Rights. 'Kenya: court rules that Lake Turkana wind power acquired community land unprocedurally', 2021, www.business-humanrights. org/en/latest-news/kenya-court-rules-that-lake-turkana-wind-power-acquired-community-land-unprocedurally/ (last accessed December 2022).

28. David Hundeyin. 'The "French Colonial Tax": a misleading heuristic for understanding Françafrique.' *The Africa Report*, 21 November 2019, www. theafricareport.com/20326/the-french-colonial-tax-a-misleading-heuristic-for-understanding-francafrique/ (last accessed December 2022).

29. 350.Org. 'African leaders shed tears and protest a "political deal"'. 350.org, 2019, https://350.org/african-leaders-shed-tears-and-protest-political-deal/ (last accessed December 2022).

30. Davies Iheamnachor. 'No oil exploration in Ogoni, until Saro-Wiwa, others are exonerated – MOSOP vows.' *Vanguard*, 9 November 2020, www.vanguardngr. com/2020/11/no-oil-exploration-in-ogoni-until-saro-wiwa-others-are-exonerated-mosop-vows/ (last accessed December 2022).

31. UNEP. 'Nigeria launches $1 billion Ogoniland clean-up and restoration programme.' UNEP, 2017, www.unep.org/news-and-stories/story/nigeria-launches-1-billion-ogoniland-clean-and-restoration-programme (last accessed December 2022).

10

Under the Yoke of Neoliberal 'Green' Trade

Rachmi Hertanti

INTRODUCTION

The US–China trade war, COVID-19 pandemic and Russia's war on Ukraine have disrupted the global supply chain of critical raw materials. This disruption is made more acute due to the significant dependence of global trade on production supplies concentrated in only a few countries. The International Energy Agency (IEA) report on the Role of Critical Minerals in the Clean Energy Transition (2021) states that the concentration of mineral production and processing in only a handful of countries makes the supply vulnerable to political instability, geopolitical risks and possible export restrictions. As shown in Figure 10.1, China, the United States and Myanmar are the top producers of rare earth elements; while Australia, Chile and China are the top producers of lithium; and Indonesia, the Philippines and Russia are the top producers of nickel.[1]

Beyond these political economic warnings, this chapter critically unpacks the geopolitical scramble for securing critical raw materials by zooming in on two interrelated dynamics: one is the insecure position of powerful countries in the global supply chain of critical raw materials, and second, the role of free trade and investment agreements. It argues that free trade and investment agreements are concrete mechanisms deployed by powerful countries, particularly those belonging to the G7 (Group of 7) countries,[2] and multinational corporations to secure the necessary minerals to produce green technology needed for a so-called green transition. A particularly insidious element of this new generation of agreements are mechanisms for investors to sue states when the formers' interests are not protected. While such mechanisms limit state sovereignty, the chapter problematises the role of the state, which is also complicit in the creation of such deals. Further, the global

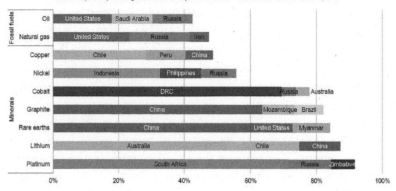

Share of top three producing countries in total production for selected minerals and fossil fuels, 2019

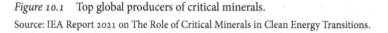

Figure 10.1   Top global producers of critical minerals.
Source: IEA Report 2021 on The Role of Critical Minerals in Clean Energy Transitions.

rush to secure critical raw materials for a green transition exposes the falla-cies of this model, that is, it is built on extracting non-renewable materials from the earth, and a 'sustainable development' agenda that does not care for the people and the planet. It reinforces a global political economic system where 'core' countries force the opening of the countries at the 'peripher-ies' as sources of critical raw materials through unfair and asymmetrical trade rules.

MORE FREE TRADE TO OPEN THE MARKET OF CRITICAL RAW MATERIAL SUPPLY CHAINS

The OECD Report on Security of Supply for Critical Raw Materials (2022)[3] presented at the G7 meeting in Germany reinforced the report of the IEA: the majority of G7 countries are not the largest producers of critical raw materials, and are not big players in the mineral processing industry for key commodities in green technology such as lithium, cobalt, bismuth and rare earth minerals. Most of the commodities are processed outside the G7 countries. The OECD further explained that the mineral reserves owned by the G7 countries are insufficient to meet their domestic industrial demand; therefore, the expansion of trade and investments in rich mineral reserves countries are needed to diversify the supply chain of critical raw materials.[4]

The G7 trade ministers agreed at a subsequent meeting in Germany in 2022[5] that they will intensify multilateral, regional and bilateral trade coop-eration to address export restrictions and trade barriers in securing critical raw materials at the international level. Some of these trade barriers pertain to addressing what they call as non-transparent and 'unfair' trade practices such as forced technology transfer, intellectual property theft, lowering of labour and environmental standards to gain competitive advantage, market-distorting actions of state-owned enterprises, and harmful industrial subsidies, including those that lead to excess capacity, all pointing to China's dominant trade practices in the supply of critical raw materials.

The G7 countries are further pushing for reforms at the World Trade Organisation (WTO), the only multilateral body that deals with trade rules between countries, to ensure that the WTO's transparency mechanism is fulfilled by all of its members in the context of the critical minerals supply chains.[6] In the WTO's transparency mechanism, export restrictions have become a significant concern and an easy target for dispute at the multilateral body. An example is the dispute on raw materials between the EU and Indonesia in 2019. The regional bloc filed a complaint at the WTO against Indonesia's policy ban on nickel exports to the EU. Indonesia implemented the restrictions in order to prioritise its domestic processing requirements and market obligations. The dispute typifies how international trade rules can be instrumentalised to ensure barrier-free global markets that facilitate unabated supply of critical raw materials for powerful countries such as the EU.

Another mechanism that G7 countries use to strengthen their supply chain resilience and contain China's dominance is fostering greater policy coherence through bilateral and transatlantic/pacific trade and investment cooperation between them and the countries that are rich in critical raw materials.[7] Currently, there are four trends of trade-related cooperation on securing supply chains of critical raw materials transpiring in the Asia-Pacific Region.

First, to counter China's growing economic influence in Asia-Pacific and the Americas, the US government initiated the Indo-Pacific Economic Framework (IPEF[8]) and the Americas Partnership for Economic Prosperity (APEP), respectively. Both instruments are designed to solve the global supply chain disruption particularly under the supply chain resilience pillar of IPEF. These will be implemented by minimising market distortions, promoting regulatory compliance, respecting market principles, and acting consistently with respective WTO obligations.[9] It is an open secret that this agreement is a trade cooperation model used by the US to strengthen its strategy of encouraging onshoring and nearshoring,[10] including significant provisions supporting only domestic supply chains as part of its inflation reduction strategy.[11]

Second, the plan to accelerate the EU Critical Raw Materials Act (CRMA)[12] to diversify trade and secure the unabated supply of critical raw materials needed for the region's green transition. Under the EU CRMA, international trade cooperation is sustained through the expansion of free trade areas and strengthening of the WTO rules[13] to ensure the regulation of unfair trade practices on export restriction measures, and enforcement of dispute settlement on trade and investment.[14] Another important element is a dedicated Energy and Raw Materials chapter in the EU FTA. Under this chapter, the EU will address raw materials-related matters such as predictable impact assessment procedures or non-discrimination treatment for investors in third countries. Currently, the EU is increasing trade cooperation with several strategic countries such as Chile, Mexico, New Zealand, Australia, India and ASEAN countries.

Third, as one of the major players in the supply chain of critical raw materials, Canada has launched the Comprehensive Economic Partnership Agreement (CEPA) negotiations with Indonesia in 2021. Canada claims that cooperation in the Asia-Pacific Region will build mutually beneficial dependencies in several vital areas of comparative advantage, namely critical raw materials, rare earth materials, agriculture, energy and natural resources.[15] But its positioning in the region is motivated by its diversification agenda, which excludes China as a source.[16]

Fourth, the ASEAN Regional Comprehensive Economic Partnership (RCEP) expands cooperation from ASEAN Plus One Free Trade Agreements (ASEAN+ 1 FTAs) to an additional five ASEAN economic partners, namely Japan, China, South Korea, Australia and New Zealand. ASEAN+1 FTAs establishes free trade areas between ASEAN countries and some of the world's major economies on a bilateral basis, in a bid to strengthen its participation in the global supply chain and increase regional integration.

RCEP is a game changer. From the beginning, RCEP was formed under the shadow of the ASEAN+3[17] cooperation to open new markets and facilitate market access in Southeast Asia. The opening of new markets, however, is not based on tariff liberalisation, but on trade facilitation mechanisms, that is, processes related to export and import of goods and services, that open up opportunities to deepen regional value chain activities through effective Rules of Origin or sets of criteria to determine the national source of a product.[18] With this set up, Rashmi Banga, a senior economist at UNCTAD stated in a 2020 study that RCEP will only have a positive impact on non-

ASEAN member states such as Japan, South Korea and China.[19] These three countries are poised to benefit more from RCEP's regional value chain scheme because they have the technology, knowhow, and are exporters of finished products such as electronic components and machineries that are sold to ASEAN countries, its main target market.[20]

MORE INVESTOR PROTECTION IN TRADE AND INVESTMENT AGREEMENTS

The expansion of investments into mineral-rich countries has consolidated activities in the mining, refining and processing raw materials sectors and green technology. Here, the role of multinational corporations (MNCs) as key investors must be foregrounded.

Mining companies are forewarned that geopolitical risks and domestic policies in mineral-rich countries will create greater investment risks.[21] To deal with these potential risks, MNCs use an extra mechanism to enable them to sue a host state that is party to an agreement. Known as Investor-State Dispute Settlement (ISDS), it is a mechanism that institutionalises corporate impunity, which strips countries' political bargaining powers to regulate corporations and protect people's rights and the environment. It has put the state as a hostage to the investors' interests. Further, foreign investors and corporations can use ISDS to seek compensation for breaches of a country's investment obligations. States would then have to pay the compensation using public monies, which consequently reduces the resources needed for social subsidies to the people. Moreover, the mechanism sends a regulatory chill that makes states complicit, failing to protect people's rights and the environment. The provision has also constricted the people's struggles against the negative impacts of corporate business activities such as human rights violations, economic losses and broader environmental damages.

This mechanism is incorporated in various international trade and investment agreements, such as the ones mentioned in the previous section. The inclusion of ISDS mechanisms in international treaties is commonly argued as a way for states to encourage foreign investment flows into a country. On the contrary, evidence shows that the mechanism has been a tool to tie the hands of a host-state in favour of protecting corporate interests and in the process, perversely shifting its bargaining power to the most powerful private economic actors, and increasing public liabilities.[22]

In the context of the climate crisis, UNCTAD has raised alarm bells over the increase of ISDS lawsuits, which either challenge (i) domestic environmental protection policies enacted to prevent and address the socio-environmental impacts of investment projects, or (ii) regulatory measures related to renewable energy production.[23] The International Centre for Settlement of Investment Disputes (ICSID)[24] report corroborates this by stressing that electric power and other energies as the top investor lawsuit cases in international arbitration court, with the oil, gas and mining sector in the second position.[25]

As the largest nickel producer country in the world, Indonesia has received lawsuit threats from foreign investors because the country has enacted a policy to restrict its export of critical raw materials. For instance, the lawsuit threats by two US companies, Newmont Nusa Tenggara[26] and Freeport McMoran,[27] challenged Indonesia's Mining Law that prioritises domestic processing requirement of raw materials over export activity. The regulatory chill effect from these investor lawsuits has undermined the state's control over natural resources.[28] Similarly, the lawsuits of Churchill Mining and Planet Mining, two corporations originally from the UK and Australia, respectively, centre on the revocation of overlapping open pit mining permits by the regional government. The mining concessions are located in conservation and protected forest areas that violated the rights of indigenous people in seven villages in Kutai Timur, Kalimantan, and damaged the environment. In this case, Churchill and Planet mining demanded a US$1 billion compensation from the Indonesian state. The same amount is equal to Indonesia's annual food subsidy allocation. The government won the case, but they still needed to pay the cost of the arbitration proceedings, which amounted to US$8 million.

BETWEEN MINERALS AND A HARD PLACE: THE CASE OF INDONESIA

With large reserves of nickel and other critical minerals needed for energy transition, Indonesia is caught in a vortex of global competition for minerals supply. Indonesia, however, plans to position itself as a finished product exporter rather than a raw material exporter in the supply chain activities for electric battery production and mineral resources. This goal is part of a green energy and industrialisation plan that intends to break away from the predatory colonialist pattern of development that heavily relies on the export of primary commodities.

To do this, the government under President Jokowi Widodo focuses on building its capacity to produce electric batteries and supplies for electric vehicles (EV),[29] as outlined in the 2019–2024 National Medium-Term Development Plan (RPJM).[30] In order to shift its position from a raw material to a finished product exporter, the government has instituted a policy to prioritise the domestic use of critical raw materials needed for the national industrialisation of EV battery production. This necessarily entailed restrictions in exports of critical raw minerals. In 2020,[31] the government banned the export of nickel ore, then bauxite in December 2022.[32] Both minerals will instead be processed and refined in the country.

According to the Indonesian government, this policy is anchored on fulfilling its constitutional mandate of using the natural wealth owned by the state for the welfare of its people.[33] It has also nationalised the mining sector based on the Mining Law 4/2009. The law requires mining companies to process and refine mining products domestically prior to export to increase the added value of mineral commodities.[34] The law intends to nationalise foreign mining companies, which compels foreign-owned mining industries to progressively divest its shares to the Indonesian government through state-owned enterprises (SOEs) or local industries, becoming a shareholder minority within five years,[35] and within ten years, increasing the shares owned by Indonesians up to 51 per cent.[36] Apart from restrictions in the export of key minerals, Indonesia is implementing protectionist policies on energy and raw materials such as limiting the prohibition on local content[37] requirements, technology transfer and the privatisation of SOEs.[38] These are designed to protect and develop domestic industries.

However, international trade rules will once again prevent Indonesia implementing its energy transition agenda. For example, EU interests under the negotiations of Indonesia–EU Comprehensive Economic Partnership Agreement (IEU–CEPA) stands to undermine Indonesia's plans. Predictably, the EU raised concerns over these protectionist policies in a scoping paper, citing that these will hinder its needed supply of critical raw materials.[39] Furthermore, the investment chapter of the IEU–CEPA adopts a similar approach to ISDS by incorporating an Investment Court System that strongly protects investors' rights and profits. It is expected that Indonesia's nationalisation agenda will provide a fodder for corporations to potentially sue the country under this bilateral agreement.[40]

Beyond the threat of powerful countries undermining the energy transition ambitions of Indonesia, the state's control over natural resources and the implementation of the current national economic transformation have raised serious concerns from people's and grassroots movements in the country. The main criticism rests on the contradictions and impacts that such extractivist and undemocratic agenda generate.

The realisation of Indonesia's economic transformation based on natural resources, particularly on battery EV industrialisation, requires an expansion of the mining and processing industry and infrastructure development on a large scale that is regulated under the National Strategic Project (PSN). In 2021, there were a total of 339 active permits for mining business[41] covering a total area spread of 836,000 hectares, which have incited and increased agrarian conflicts. Around 40 conflicts involving 11,466.923 hectares have broken out.[42] These conflicts are often characterised by human rights abuses and socio-environmental impacts affecting numerous communities.

Further, Indonesia's industrialisation plan still heavily relies on foreign investments. The government requires USD 30.9 billion as a total investment pipeline for the development of the country's electric vehicle battery supply chain industry.[43] In order to attract foreign investment to support this industrial policy, the government passed a controversial national regulation act called the Omnibus Law on Job Creation that makes it easy for foreign investors to obtain business licences, including permits to access natural resources and land lease, and relax labour laws.[44]

The contradictions in the Indonesian government's energy transition call into question the role of the state in realising just transition and transformation in the country. For this author, just transition comprises a systemic turn, through genuinely democratic means, away from exploitation, extraction and alienation, towards systems of production and reproduction that are focused on human well-being and the regeneration of ecosystems.[45] Just transition is much more than a shift from fossil fuels to renewable or green energy sources.[46]

But even within this issue of energy transition, state and corporate ownerships and control and access to energy sources, materials and necessary technologies must be questioned. This also necessitates challenging free trade and investment agreements that limit the possibility of democratising energy transition for the people and the planet. It entails shifting the power and control in all aspects of the sector – from production to distribution and supply, finance to technology and knowledge, and energy users and workers.[47] In the case of Indonesia, this means questioning the state-led energy transition agenda and its continued reliance on trade rules and agreements that paradoxically undermines this very agenda.

CONCLUSION

The current dynamics in the critical minerals supply race bring to the fore political economic questions inherent in the green energy transition agenda. Can extraction for the energy transition be done in a socially and environmentally just manner? Is extraction without extractivism even possible? In the context of increased geopolitical and geoeconomic rivalry, what (transformative) role is there for the state? Can capital be disciplined through public and democratic ownership of natural resources? Or is this just wishful thinking?

While these questions are beyond the scope of this chapter, these questions fundamentally require a rethinking of just transition. A truly just transition means embarking on a process that develops the praxis of a people's energy transition led by working-class peoples in the country.[48] A people's energy transition must be based on constructing the right to energy, strengthening diverse forms of the public sphere, decentralising and democratising decision-making processes around energy, and constructing new social relations anchored on the harmony between people and Nature.

NOTES

1. International Energy Agency. 'The role of critical minerals in clean energy transitions'. *OECD Publishing* (2021), 287.
2. G7 is an intergovernmental political forum comprising Canada, France, Germany, Italy, Japan, the United Kingdom and the United States; additionally, the European Union is a 'non-enumerated member'.
3. G7 Leaders' Summit. 'Security of supply for critical raw materials: vulnerabilities and areas for G7 coordination'. Schloss Elmau, 26 June 2022, OECD_2022-07-01-Security-of-Supply-for-Critical-Raw-Materials-Data.Pdf, n.d.
4. It reinforces the output of G7 meeting in 2021 held in Cornwall, UK, hammered out an agenda called 'Build Back Better for the World (B3W)'. The agenda of B3W was proposed as a G7's global recovery strategy in addressing the global infrastructure gap, COVID-19 recovery, and decouple China's dominance and influence in the developing world ('The Role of the G7 in Mobilizing for a Global Recovery', Chatham House).
5. G7. 'G7 trade ministers' statement Neuhardenberg'. Federal Ministry for Economic Affairs and Climate Action, 15 September 2022, 20220915-G7-Trade-Ministers-Statement-Neuhardenberg-15-September-2022.Pdf.
6. 'G7 trade ministers eye progress on WTO reform, economic coercion', InsideTrade.Com', https://insidetrade.com/daily-news/g7-trade-ministers-eye-progress-wto-reform-economic-coercion (last accessed 8 November 2022).

7. 'US and European strategies for resilient supply chains.' Chatham House – International Affairs Think Tank, 14 September 2021, www.chathamhouse. org/2021/09/us-and-european-strategies-resilient-supply-chains.

8. The IPEF members are Australia, Brunei Darussalam, Fiji, India, Indonesia, Japan, the Republic of Korea, Malaysia, New Zealand, the Philippines, Singapore, Thailand and Vietnam.

9. Indo-Pacific Economic Framework for Prosperity (IPEF). 'Ministerial statement for Pillar II of the Indo-Pacific Economic Framework for Prosperity.' Commerce, www.commerce.gov/sites/default/files/2022-09/Pillar-II-Ministerial-Statement.pdf (last accessed 2 April 2023).

10. Onshoring refers to the relocation or establishment of production facilities within a country's borders. Nearshoring, on the other hand, involves setting up facilities in close proximity to the country of demand or consumption. In the context of the critical raw materials supply chains, both strategies aim to spur domestic control over supply and reduce dependence on foreign suppliers, and in the process stimulating both industrialisation and reduction of inflation rates.

11. Joseph Majkut. 'The Inflation Reduction Act: a race to the top or protectionism in high gear?', Center For Strategic & International Studies (CSIS), 1 March 2023, www. csis.org/analysis/inflation-reduction-act-race-top-or-protectionism-high-gear.

12. The following raw materials shall be considered strategic for the EU under the Critical Raw Materials Act: bismuth; boron – metallurgy grade; cobalt; copper; gallium; germanium; lithium – battery grade; magnesium metal; manganese – battery grade; natural graphite – battery grade; nickel – battery grade; platinum group metals; rare earth elements for magnets (Nd, Pr, Tb, Dy, Gd, Sm, and Ce); silicon metal; titanium metal; tungsten.

13. European Commission (EC). 'European Critical Raw Materials Act', https:// ec.europa.eu/commission/presscorner/detail/en/ip_23_1661 (last accessed 30 March 2023).

14. 'Critical raw materials: ensuring secure and sustainable supply chains for EU's green and digital future', European Commission, https://ec.europa.eu/commission/presscorner/detail/en/ip_23_1661 (last accessed 30 March 2023).

15. Stephen R. Nagy. 'Canada critical to helping ensure global supply chain security, The Japan Times, 15 February 2022, www.japantimes.co.jp/opinion/2022/02/15/commentary/world-commentary/canada-supply-chains/.

16. 'EU and Canada set up a strategic partnership on raw materials.' European Commission, https://single-market-economy.ec.europa.eu/news/eu-and-canada-set-strategic-partnership-raw-materials-2021-06-21_en (last accessed 8 November 2022).

17. Plus three refers to Japan, China and South Korea.

18. 'Technical information on rules of origin', World Trade Organization, www.wto. org/english/tratop_e/roi_e/roi_info_e.htm.

19. 'RCEP: goods market access implications for ASEAN', Global Development Policy Center, 24 March 2021, www.bu.edu/gdp/2021/03/24/rcep-goods-market-access-implications-for-asean/.

20. Press Release: ASEAN's fate hangs in an RCEP balance', APWLD, 25 March 2021, https://apwld.org/press-release-aseans-fate-hangs-in-an-rcep-balance/.
21. Paul A. Bendall. 'Mine 2022: a critical transition', PwC, 2022, p. 36.
22. Gus Van Harten. 'A critique of investment treaties'. *Rethinking Bilateral Investment Treaties: Critical Issues and Policy Choices* (2016): 41–50.
23. United Nations UNCTAD. 'IIA issues note', International Investments Agreements, No. 4 (2022): 22.
24. As part of the World Bank Group, ICSID is an international arbitration institution established in 1966 for legal dispute resolution and conciliation between international investors and states.
25. 'Annual Report 2022', International Centre for Settlement of Investment Disputes, https://icsid.worldbank.org/sites/default/files/publications/ICSID_AR.EN.pdf (last accessed 8 November 2022).
26. Hilde Van Der Pas and Riza Damanik. 'The case of Newmont Mining vs Indonesia', TNI, 12 November 2014, www.tni.org/en/publication/the-case-of-newmont-mining-vs-indonesia.
27. Iwan Supriyatna. 'Pemerintah Diminta Cuek terhadap Ancaman Freeport', Kompascom, 21 February 2017, https://money.kompas.com/read/2017/02/21/112522526/NaN
28. 'Gugatan ISDS: Ketika Korporasi Mengabaikan Kedaulatan Negara', Indonesia for Global Justice, 2019, https://igj.or.id/wp-content/uploads/2019/10/Majalah-IGJ-ISDS-Lawsuit-compressed.pdf.
29. Humas. 'Inilah Perpres No. 55/2019 tentang Program Kendaraan Bermotor Listrik Berbasis Baterai'. Sekretariat Kabinet Republik Indonesia, 15 August 2019, https://setkab.go.id/inilah-perpres-no-55-2019-tentang-program-kendaraan-bermotor-listrik-berbasis-baterai/.
30. 'Lampiran Peraturan Presiden Republik Indonesia'. Narasi, 18 (2020), https://perpustakaan.bappenas.go.id/e-library/file_upload/koleksi/migrasi-data-publikasi/file/RP_RKP/Dokumen%20RPJMN%202020-2024/Lampiran%20 1.%20Narasi%20RPJMN%202020-2024.pdf (last accessed 22 February 2023).
31. 'Bijih Nikel Tidak Boleh Diekspor Lagi per Januari 2020', ESDM, January 2020, www.esdm.go.id/id/media-center/arsip-berita/bijih-nikel-tidak-boleh-diekspor-lagi-per-januari-2020 (last accessed 23 March 2023).
32. Office of Assistant to Deputy Cabinet Secretary for State Documents & Translation. 'Gov't to impose export ban on bauxite ore June next year'. Sekretariat Kabinet Republik Indonesia, 21 December 2022, https://setkab.go.id/en/govt-to-impose-export-ban-on-bauxite-ore-june-next-year/.
33. Article 33 paragraphs (2) and (3) of the 1945 Constitution. Paragraph (2) states: 'Sectors of production which are important for the state and which affect the life of the people at large are controlled by the state.' Paragraph (3) states: 'Earth, water, and the natural resources contained therein are controlled by the state and used for the greatest prosperity of the people.'
34. Article 102 and 103 of the Mining Law 4/2009. Article 102 states: 'IUP and IUPK holders are required to increase the added value of mineral and/or coal resources in the implementation of mining, processing and refining, as well as utilization minerals and coal.' Article 103 states: 'Holders of IUP and IUPK

Production Operations are required to process and refine mining products domestically.'

35. Article 112 of the Mining Law 4/2009 states: 'After 5 (five) years of production, business entities holding IUP and IUPK whose shares are owned by foreigners are required to divest their shares to the Government, regional government, state-owned enterprises, regionally-owned enterprises, or national private enterprises.'

36. Article 27 of the Regulation of the Minister of Energy and Mineral Resources 25/2018 states: 'the IUP Holders of Production Operation and IUPK Holders of Production Operation in the framework of foreign investment, since 5 (five) years after production are required to carry out a gradual divestment of shares, so that in the tenth year at least 51% (fifty one percent) of the shares are owned by Indonesian participants.' The Mining Law 4/2009 has been amended to new Mining Law No.3/2020, but similar obligations and requirements were reaffirmed in it.

37. Local content requirements are rules that compel a foreign company to source a certain amount of its final good or service from domestic firms, either through local purchases or actual production.

38. 'Kemenperin: Pemerintah Targetkan Indonesia Jadi Pusat Kendaraan Listrik Di ASEAN', Kemenperin, www.kemenperin.go.id/artikel/21279/ghs (last accessed 31 March 2023).

39. See the section of Raw Materials and Energy in the Scoping Paper CEPA of Indonesia-EU, 16.

40. 'The EU–Indonesia CEPA negotiations.' Transnational Institute, 1 February 2023, www.tni.org/en/publication/the-eu-indonesia-cepa-negotiations.

41. Direktorat Jenderal Mineral dan Batubara. 'Grand Strategy Mineral Dan Batubara.' ESDM, 2021, www.esdm.go.id/assets/media/content/content-buku-grand-strategy-komoditas-minerba.pdf (last accessed 26 February 2023).

42. Konsorsium Pembaruan Agraria. 'Menuntun Jalan Sejati Reforma Agraria.' Menuntun Jalan Sejati Reforma Agraria, 24 September 2022, www.kpa.or.id/kabar-agraria/view/menuntun-jalan-sejati-reforma-agraria_38b3eff8baf56627478ec76a704e9b52 (last accessed 31 March 2023).

43. 'Menkomarves Catat Pipeline Investasi Hilirisasi Tambang US$ 30,9 Miliar, Berikut Daftarnya.' investor.id, https://investor.id/business/315583/menkomarves-catat-pipeline-investasi-hilirisasi-tambang-us-309-miliar-berikut-daftarnya (last accessed 31 March 2023).

44. 'RUU Omnibus Cipta Lapangan Kerja.' Indonesia for Global Justice, April 2020, https://igj.or.id/wp-content/uploads/2020/04/Framing-Paper-IGJ_RUU-Omnibus-Cilaka.pdf (last accessed 15 March 2023).

45. See www.tni.org/en/publication/just-transition.

46. Transnational Institute. 'From crisis to transformation', 1 February 2023, www.tni.org/en/publication/from-crisis-to-transformation.

47. Transnational Institute. 'Towards energy democracy', 1 February 2023, www.tni.org/en/publication/towards-energy-democracy.

48. Transnational Institute. 'Towards a corporate or a people's energy transition?', 1 February 2023, www.tni.org/en/publication/towards-a-corporate-or-a-peoples-energy-transition.

11

'Nature-Based Solutions' for a Profit-Based Global Environmental Governance[1]

Mary Ann Manahan

INTRODUCTION

The multilateral system is the domain most often expected to take on the crucial task of directly addressing the complexities of eco-social transformation. Global agreements and instruments such as the UN Framework Convention on Climate Change and the Convention on Biological Diversity are supposed to be linchpins to navigate the path towards a 'sustainable' future. However, the multilateral system is in crisis.[2] Specifically, multilateralism, as primarily embodied by the United Nations (UN) system, has been marred by contestations around its legitimacy, credibility and relevance as a global governance system that can foster international cooperation and action around the most pressing issues of the time, such as climate change, growing inequality, health and the COVID-19 pandemic. But the crisis of multilateralism is hardly new. It is but a continuation of politico-historical processes that began in the late 1970s.

According to Ethiopian-American political scientist Adom Getachew, this decade was marked by a neoliberal counterrevolution of capital that brought significant changes within the UN system. First, the rejection of the New Economic International Order (NIEO)'s vision of dismantling the economic vestiges of colonialism.[3] The NIEO – adopted by the UN General Assembly in 1974 – was championed by newly independent countries from the Global South that advocated for international welfare, redistribution and global justice through a new interdependent global system as pathways to end exploitative and unfair trade relations and a Western-built financial order that structurally maintained US global hegemony.[4] Second, the gradual American withdrawal from the UN and other multilateral institutions in the post-Second World War period.[5] Rather than engage in a rules-based international order, the United States (US) embraced 'a new sovereigntism' that advanced military intervention and rejected international norms.[6]

Third, the critique of public inefficiency and state regulation as an unjust restriction on market freedom became pervasive. Proponents of this critique claimed that the private sector and market were more efficient, and therefore should have an increased role and space in public policy and governance. At the beginning of the 1980s, this neoliberal thought became hegemonic, imposing managerial corporate logics of 'quality' and 'efficiency' as universal standards that public institutions across multiple scales must adopt.[7] This development was further reinforced through the World Bank's popularisation of the concept of 'governance' in the 1990s that tacitly opened the space for corporations as part of 'civil society' to participate in decision-making on a slew of international topics, especially in the provisioning of global public goods.[8]

Multistakeholderism became the new approach to implement this paradigmatic shift that introduced 'a whole new set of governance actors and a new process for making global laws and regulations'.[9] The twenty-first century witnessed the rising influence of transnational corporations (TNCs) and other private actors that have vested political and economic interests over issues of public interest and the fate of the planet. Their roles were further galvanised through the modality of private–public partnerships[10] and the UN Global Compact, which is the largest corporate 'sustainability' initiative in the 2000s.[11]

With its roots in corporate management science and practice, *stakeholders* in multistakeholderism refers to organisations and individuals that have a 'stake' or an interest in discussing a specific policy challenge or generating actions to address it. However, the language of 'stakeholders' and 'stake', especially at the global level, is imbued with contested political issues and power asymmetries that are often masked or side-stepped in the attempt to reach action-driven consensus. Not all *stakeholders* hold equal positions and not all *stakes* define the agenda, plans and actions within multistakeholder initiatives. Often, those that are more powerful – be it countries, donors or corporations – get to tip the scales of governance.

This chapter offers a critique of multistakeholderism as a form of privatised global governance marked by corporate capture, a deficit of democracy and accountability, and a complicit UN. It contributes to the debates on the geopolitics of eco-social transformation and global (in)justice by critically exploring multistakeholderism's concrete articulations in global environmental governance. It looks into current entanglements between the geopolitical Norths and Souths, arguing that the global environmental governance under multistakeholderism is undergirded by colonial and capitalist discourses and perspectives about Nature. These epistemes gave rise to so-called 'nature-based solutions', which big polluting TNCs and other private actors are lucratively investing in. Driven by the neoliberal logic that a solution is only viable if it is, first and foremost, profitable, multistakeholder initiatives around 'nature-based solutions' are predominantly shaping the boundaries of solutions to the eco-social crises. Finally, the chapter concludes by reflecting on the possibilities of reconstituting radical democratic multilateralism under the current global governance system.

THE PRIVATISATION OF GLOBAL ENVIRONMENTAL GOVERNANCE

Several coalescing dynamics facilitated the rise of multistakeholderism in the governance of global ecosystems. One is the exercise of national political authority by powerful, pollution-producing governments regarding eco-social issues that would require collective action beyond national boundaries. An example is the insistence of the US and the EU to consider the rate of greenhouse gas emissions as a national matter, and not an issue of decision-making at the international level.[12] Such blatant display of national rule by powerful countries when it suits their interests has stymied multilateral environmental governance.

Second is the institutionalisation of 'sustainable development', a term which was placed on the map through the 1987 'Our Common Future' Report and the 1992 Earth Summit that engendered Agenda 21 and the Rio Declaration on Environment and Development. Sustainable development was defined in the Brundtland report as 'development that meets the needs of the present without compromising the ability of future generations to meet their own needs'.[13] While the concept has been largely debated for its ambiguity, at the core of it is the affirmation that limitless economic growth and protection of Nature can go together, even on a finite planet, and that there is no inevitable conflict between the two.[14] Twenty years later, at the Rio+20 Summit, the same developmental goals of profit, people and planet were renewed by UN member states. But this time even advocating for a 'mutually reinforcing... relationship of economic growth, nature protection and social equity objectives'[15] under a new frame – the 'green economy' – at the heart of which is a continuation of the dominant logic of depredatory, neoliberal capitalism.

Under this new framing, saving Nature and biodiversity can only be achieved through the re-valuing and incorporation of peoples and ecosystems into the terms set by financial and global markets.[16] This has meant refashioning environmental governance efforts and actions by states, the private sector, and civil society moving from centralised, state-led environmental regulations to devolved, market-oriented and -based approaches (e.g. payment for ecosystem services).[17] The reorientation to market-based approaches is deemed as the 'alternative' response to heavy and strict regulations in the 1960s/1970s, or what Naomi Klein refers to as the 'golden age of environmental legislation', underpinned by the 'polluters pay' principle and severely limiting pollution and other forms of environmental degradation through effective sanctions.[18] According to Klein, these policies were designed to regulate pollution and protect public health and the environment, but were then heavily criticised by neoliberals as costly, inefficient and oppressive, arguing that they impede economic growth and innovation.[19] In other words, neoliberals accused these environmental regulations as being burdensome for businesses, and instead, argued for voluntary approaches to environmental regulations and market-based solutions as more efficient ways to address socio-ecological problems.

The third and last dynamic concerns the dwindling resources coming from the UN's wealthier members, particularly the US, which led to an increasing reliance of UN agencies on mega-philanthropies and corporate donors to fill this funding gap. Since the terms of former UN Secretaries-General Boutros Boutros-Ghali and Kofi Annan,[20] this close alliance between the UN and corporate actors was further entrenched through various global agreements such as the Millennium Development Goals (MDGs), Sustainable Development Goals (SDGs) and the 2015 Paris Agreement.[21] The latest agreement that formalised the relations between the UN and the World Economic Forum (WEF), a global association of the most influential individuals and institutions from business, politics, science and culture, is the 2019 Strategic Partnership Framework aimed at deepening institutional arrangements to accelerate the SDG implementation.[22] This agreement legitimises what WEF founder and director Klaus Schwab calls 'stakeholder capitalism', or 'a new stakeholder paradigm of international governance analogous to that embodied in the stakeholder theory of corporate governance on which the [WEF] itself was founded'.[23] This brand of multistakeholderism reflects a capitalist discourse that prioritises the interests of corporations and shareholders, while it seemingly calls for a broader stakeholder approach to global governance. The problem here is not only that the WEF is a private club of the rich and wealthy without any democratic legitimacy, but also that it unabashedly advocates for a model of governance that is technocratic and elitist in approach, disconnected from the needs and voices of local communities and majority of the world's population. Therefore, it is highly dubious that its call for a paradigm shift in governance will lead to more equitable and just outcomes for all 'stakeholders'.

A mapping exercise of twenty-six multistakeholder initiatives[24] revealed the UN's complicity and high degree of influence wielded by business and big NGOs in these spaces. For example, corporations dominate in the decision-making structures of the Renewable Energy and Energy Efficiency Partnership (REEEP). Originally launched by the United Kingdom government in 2002 at the Johannesburg World Summit on Sustainable Development, REEEP aimed at facilitating market transformation for renewable energy in Asia and Africa using private and public funding mechanisms. As an autonomous entity backed by the United Nations Industrial Development Organisation, its governing and advisory boards consist of large energy (e.g. General Electric, First Energy Asia and Enel) and investments companies (e.g. Glennmont Partners, Southbridge Investments and Finite Carbon) that have accessed emerging markets in the Global South.

UNPACKING GLOBAL ENVIRONMENTAL MULTISTAKEHOLDERISM

In the same mapping exercise mentioned above, two distinct thematic foci and discourse imperatives of environmental multistakeholder initiatives were identified. From 2000 to 2010, a majority of them focused on managing forest use, regulating the minerals, oil and gas sectors, as well as on newly emerging biofuel industries, and securitising the environment. All these topics have advanced a framing that the environment, biodiversity and Nature are extremely and existentially threatened, and because of the urgency, political debates and democratic deliberation about responses can be justifiably bypassed.[25] This framing facilitated the construction of social and environmental standards that heavily rely on corporate social responsibility and the voluntary compliance of companies, reducing governmental action to monitoring.

The focus on voluntary standards engendered multiple roundtables focused on export commodities (e.g. palm oil, diamonds) and eco-labels (e.g. Marine Stewardship Council). For example, in 2002, the Extractive Industry Transparency Initiative (EITI), a multistakeholder global standard, was formed to promote open and accountable management of oil, gas and mineral resources in 55 countries. It was backed by a coalition of governments, NGOs and businesses. However, its rhetoric of accountability and transparency obscures EITI's neglect of the participating governments' corruption and the socio-environmental and human rights impacts of the extractive industries.[26] EITI has been a tool for state and corporate green-washing, as well as the expansion of new territories for mining.

From 2011 to 2020, new(er) themes were covered by multistakeholder initiatives such as climate finance, nature-based and natural climate solutions and renewable energy. These initiatives were formed during a period of intense multilateral discussions and negotiations on climate change at the global stage, on one hand. On the other, there was an increased global awareness and demands for action to address the effects of the 'Anthropocene', a new unofficial geological period to mark the undeniable impacts of human activities on the Earth's climate and ecosystems.[27] Nature-based solutions and their iterations such as natural climate solutions, nature-positive and forest-positive future, emerged as another set of sustainability buzz-words. Their strategies underscore the idea of 'working with nature to do what it is already doing for millions of years: sequester and store carbon'.[28] They cover conservation, restoration and land-based mitigation efforts that seek to increase carbon storage and/or prevent greenhouse gas emissions in forests, landscapes and wetlands across the world. In a corporate propaganda video,[29] major emitters and polluters such as Shell unabashedly call for businesses to stand together to unlock the (profit) potential of nature-based solutions in addressing the climate crisis.

Concrete examples are the WEF-led multistakeholder initiatives called the Natural Climate Solutions Alliance (NCSA) and Nature for Climate (Nature4Climate), whose primary objectives are to increase investments and influence policymakers to induce actions, particularly in the agriculture and forestry sectors. The initiatives advocate for the creation of new metrics and business models that capitalise on the transitions needed in three socio-economic systems[30] that, according to the proponents, can ostensibly deliver EUR9.3 trillion of annual business opportunities and 395 million jobs by 2030.[31] The WEF lists 'innovative technology-driven' business models such as alternative proteins to food and waste saving technologies. The WEF emphasises a narrative that huge profits can be made amidst the climate and ecological crisis if corporations are willing to shift their operations and transition to not only sustainable, but 'nature-positive' practices.

What are the main problems with these new sustainability frames? They reinforce exploitative Nature–society relations that have brought the destruction of the basis of life in the first place. That means, they will not be effective in building environmental sustainability. Further, euphemisms such as nature-based solutions are the latest conservation hype being used to push for various forms of offsets and techno-fixes in the Global South. Carbon offset projects, for instance, have often induced massive land grabbing, displacements and dispossessions, human rights abuses against indigenous and forest-based communities, while also allowing for pollution on the side of the industries interested in offsetting.[32] Nature-based solutions, therefore, are new forms of greenwashing that perversely incentivise big corporate polluters to continue profiting from plundering the earth and causing human misery.

NATURE AS AN ASSET FOR ACCUMULATION STRATEGIES

Multistakeholder initiatives around nature-based solutions entrench the valuation of Nature as capital, an economic asset that fundamentally has a price tag on all its dimensions, services and functions (e.g. water purification by pristine watersheds or carbon sequestration of forests and oceans).[33] It consequently draws Nature into financialised markets, simultaneously locking ecosystems into the boom and bust of the financial world, as well as dislocating forests, oceans, and lands from their places of origin, histories, relations with people and communities that rely on them.[34]

Nature's valuation as an economic asset is best articulated by the Natural Capital Declaration that seeks to develop models to monetise, marketise and commodify Nature and the services it provides.[35] Signed by top CEOs of various global corporations with the support of UNEP during the Rio+20 Summit in 2012, it contends that:

Natural Capital comprises Earth's natural assets (soil, air, water, flora and fauna), and the ecosystem services resulting from them, which make human life possible. Ecosystem goods and services from Natural Capital are worth trillions of US dollars per year and constitute food, fiber, water, health, energy, climate security and other essential services for everyone. Neither these services, nor the stock of Natural Capital that provides them, are adequately valued compared to social and financial capital. Despite being fundamental to our wellbeing, their daily use remains almost undetected within our economic system. Using Natural Capital this way is not sustainable. The private sector, governments, all of us, must increasingly understand and account for our use of Natural Capital and recognize the true cost of economic growth and sustaining human wellbeing today and into the future.[36]

This declaration signals the corporate and financial sector's 'commitment' to work towards integrating 'natural capital' into their visions, strategies, operations, products and services. It ushers in the capitalist invasion into Nature that values 17 ecosystem services and 16 biomes in economic terms, worth at least EUR15–50 trillion.[37] The declaration also birthed various multistakeholder initiatives such as the Natural Capital Coalition and Capitals Coalition.

However, valuing nature as capital for accumulation strategies have far-reaching implications. First, Nature can only be saved if a price tag is put on it. This has generated new markets such as the Ecosystems Marketplace and cap-and-trade, and it also opens the possibility for the depletion of Nature at the very moment this becomes more profitable than its conservation. Second, it requires new modalities, global rules and decision-making infrastructures that facilitate the involvement of corporate and financial actors to push for its mainstreaming at multiple governance levels. An example is the creation of the Natural Capital Protocol,[38] a new standardised framework for business to measure, manage and value their impacts and dependencies on Nature. Third, this transformation of our understanding of Nature extends to changes in society–Nature relations. For instance, payment for ecosystem services espoused by The Economics of Ecosystem and Biodiversity, a global initiative to make Nature's economic values visible, has transformed indigenous peoples and forest-based communities into ecosystem service sellers and providers, and urban residents, industries/corporations, etc. into users and buyers of these services, reducing complex rural–urban relations to simple financial transactions. This episteme, therefore, reshapes the lived realities of peoples on the ground.

Further, the pervasiveness of this capitalist discourse on Nature has been made possible through interlocking strategies deployed by multistakeholder initiatives: combining convenorship with 'scientific' knowledge production and creating epistemic communities. Lead organisations coming from the corporate sector connect with big international NGOs, academic and research community and UN agencies to synergistically disseminate their narratives and solutions to environmental issues they deem as ungoverned or inadequately addressed.

To provide an illustrative example, the Natural Capital Coalition uses this strategy to advance the idea of 'natural capital' by bringing together more than 300 governmental, business, selected conservation organisations, mega-philanthropies, aid agencies and UN agencies to support the development of methods for natural capital valuation in business. This, then, engenders well-connected, self-referential networks or epistemic communities,[39] which are often viewed to have 'recogni(z)ed expertise and competence in a particular domain and an authoritative claim to policy-relevant knowledge within that domain or issue-area'.[40] The huge complexity of these structures, their economic, political and institutional power and their claim to know make it difficult for grassroot actors to oppose their strategies.

NORTH–SOUTH ENTANGLEMENTS

'Nature-based solutions' are not only aggressively disseminating capitalist logics. They are also imbued with colonial and cultural domination or what decolonial scholars, Gurminder Bhambra and Peter Newell call as 'colonialism by corporations', a contemporary phenomenon in which TNCs' plunder natural resources and exploit labour in the Global South, perpetuating historical patterns of colonialism.[41] An example is the Roundtable on Sustainable Palm Oil (RSPO), a multistakeholder governance body that provides voluntary certifications to promote the growth and sustainability of the palm oil industry. The monoculture practice of oil palm plantations is a colonial invention that primarily benefited the colonial authorities and the metropoles.[42] Oil palm plantations have also been associated with subjugation, land dispossession and labour exploitation of colonial subjects. These colonial practices are still continuing today despite the oxymoronic claims of sustainable palm oil by RSPO. In fact, it has been sternly criticised for certifying several industrial palm oil companies that destroy tropical forests, displace local populations, induce biodiversity loss, foment land conflicts and violate peoples' rights in Africa and Asia, despite community grievances and opposition.[43] RSPO's certification scheme is well-suited for large-scale plantations that prioritise productivity, technology and substantial financial resources, which entrench corporate power.[44]

Further, transnational conservation organisations that are active in multistakeholder initiatives advance nature-based solutions in alliance with TNCs and states by evoking renewed calls for the creation of protected areas. Tropes on the protection of forests as the 'last frontier' invoke colonial and romanticised constructs about Nature and wilderness captured by the *Terra Nullius* doctrine *(of vast empty, uninhabited lands)*.[45] Efforts to include indigenous peoples as 'natural partners' in conservation are imbued with common tropes of blaming them for environmental degradation because they have lost their cultural values and traditional practices of relating with Nature and forests, and therefore, the solution is to restore their traditional roles through education performed by non-indigenous, often Western conservationists.[46]

These points relate to Global South countries and racialised peoples' '(in)capacities' to protect the environment. Getachew argues that colonialism disrupted existing political and social structures of pre-colonial societies, often replaced with institutions that served the interests of colonial powers.[47] This has greatly weakened Southern societies' capacity to self-govern and create their own institutions. In the context of multistakeholderism, global platforms were set up to promote 'responsible' production of internationally traded resources with the view that Global South countries which are host to these resources lack the capacity to set their own socio-environmental standards, and that Northern environmental NGOs, governments and corporations should, therefore, define what constitutes a responsible resource management. In these global certification systems, the Global South countries have limited influence in shaping the standards and decision-making processes.

Another aspect of the North–South entanglements is resource extraction. This is perhaps best typified by the Kimberly Process Certification Scheme (KPCS), a policy-oriented multistakeholder initiative that aims to prevent the trade in conflict diamonds that are mined in war zones and used to finance armed conflicts and human rights abuses. Critical scholars and activists have decried the scheme for entrenching unequal power relations as it provides Northern countries and corporations the power to define the standards for what can be considered as 'conflict diamonds'. It also perpetuates white saviourism by positioning these powerful actors from the North as the saviours of Southern mining communities from conflict diamonds.

Equally insidious is how those schemes overlook the fact that until present times, resource extraction and labour right abuses in the Global South are often driven by Northern demands for luxury goods and minerals as well as by the unjust global rules of trade and financial institutions. At the same time, peripheral states and their political-economic elites create niche opportunities that deploy corrupt and rent-seeking practices because they have always served to channel riches to foreign global elites (see Hertanti's and Bassey's chapters in this book). Global South states are not innocent victims of these North–South entanglements, but active players that reproduce the colonial matrix of the capitalist world system. In this colonial matrix, multistakeholder initiatives continue to perpetuate profoundly asymmetric core–periphery relations.

RECONSTITUTING RADICAL MULTILATERALISM?

How do we get out of this quagmire? How do we reimagine and reconstitute a global governance that curbs the power of corporations, core-industrialised countries and other dominant actors in the Global South, and that puts primacy on the voices and realities of the governed? To reimagine the future of a new radical multilateralism, the past provides some guideposts.

The NIEO demanded a code of conduct for TNCs as a pillar of promoting economic justice and decolonisation in the Global South. It was aimed at strictly regulating the activities of corporations and ensuring that they respect the sovereignty and rights of communities and countries where they operate. Such demands also reverberate in contemporary global campaigns led by social movements for an internationally binding treaty to regulate the activities of TNCs and other businesses.[48] A growing counter-hegemonic epistemic community comprising social movements, progressive NGOs, scholars and research institutions are putting forward proposals to counter the corporate takeover of global governance.

One is the introduction of a UN-wide Corporate Accountability Framework proposed by a panel of international experts on the food system, called IPES-Food. The proposal reinforces calls for a legally binding instrument that requires TNCs to conduct human rights and environmental due diligence in their global operations and provide affected communities with access to justice and remediation.[49] Envisioned to be enforced by an independent international body, it veers away from problematic voluntary compliance mechanisms for corporations in international human rights law. However, for such a regulation to become possible, existing power relations within multilateral bodies and conflicts of interest must urgently be acknowledged and addressed, instead of perpetuating the diplomatic narrative of a harmonious community of nations and stakeholders that all pursue the same goals (see also Brand and Lang's chapter).

Beyond these demands to reform the current multilateral system, reconstituting multilateralism requires recentring calls for the redistribution of resources, wealth and decision-making power that not only rectify historical injustices, but also create a radically just multilateral system that foregrounds the needs and aspirations of marginalised communities worldwide. This includes seriously pursuing initiatives and campaigns for reparations that demand just compensation, both monetary and non-monetary, for ecological damages by extractive industries and rich countries, which can be redirected to local and indigenous communities' self-determination projects.

Moreover, new forms of multilateralism require building new projects of governance anchored in the principles of solidarity and mutual aid rather than competition and individualism. The governed – generally targets of global governance such as grassroots communities and social movements – can take the lead in these spaces. To construct new forms of collaboration entails grappling with and addressing the legacies of colonialism that continue to shape global power relations and prevent the realisation of a democratic multilateralism. This task is certainly complex and challenging; but one that requires deep commitment to social justice and socio-ecological transformation and the creation of global rules that allow different pathways of world-making and sustaining life to thrive.

NOTES

1. This chapter draws from and expands on the author's published work, titled *The Great Takeover: Mapping of Multistakeholderism in Global Governance*, written with Madhuresh Kumar (Transnational Institute, 2021).
2. Sofia Monsalve. 'Re-grounding human rights as cornerstone of emancipatory democratic governance'. *Development* 64 (2021): 13–18.
3. Adom Getachew. *Worldmaking After Empire: The Rise and Fall of Self-Determination* (Princeton, NJ: Princeton University Press, 2019).
4. Malick Doucouré (2023). 'The new international economic order', www.newagebd.net/article/198941/the-new-international-economic-order (last accessed 10 May 2023).
5. Getachew, *Worldmaking*, 176–77.
6. Ibid., 178–79.
7. Edgardo Lander and Santiago Arconada Rodriguez. *Crisis Civilizatoria: Experiencias de Los Gobiernos Progresistas y Debates En La Izquierda Latinoamericana* (Bielefeld, Germany: transcript Verlag, 2019).
8. See Inge Kaul, Isabelle Grunberg and Marc Stern (eds). *Global Public Goods: International Cooperation in the 21st Century* (Oxford: Oxford University Press, 1999).
9. Harris Gleckman. *Multistakeholder Governance and Democracy: A Global Challenge* (London: Routledge, 1998), p. 37.

10. Nora McKeon. 'Are equity and sustainability a likely outcome when foxes and chickens share the same coop? Critiquing the concept of multistakeholder governance of food security.' *Globalizations* 14(3) (2017): 379–98, DOI: 10.1080/14747731.2017.1286168.

11. Mary Ann Manahan and Madhuresh Kumar. *The Great Takeover: Mapping of Multistakeholderism in Global Governance* (Amsterdam: People's Working Group on Multistakeholderism, 2021).

12. Gleckman, *Multistakeholderism*.

13. United Nations. 'Brundtland Commission Report: Our Common Future', 1987, www.un-documents.net/our-common-future.pdf (last accessed 15 April 2023), 14.

14. SDG Knowledge Platform. 'About the Rio+20 Conference', https://sustainable development.un.org/rio20/about (last accessed 15 April 2023).

15. Peter Wilshusen. 'Capitalizing conservation/development'. *Nature Inc.* (2014): 19.

16. Bram Büscher, Sian Sullivan, Katja Neves, Jim Igoe and Dan Brockington. 'Towards a synthesized critique of neoliberal biodiversity conservation'. *Capitalism Nature Socialism* 23(2) (2012): 4–30.

17. Wolfram Dressler, Melanie McDermott, Will Smith and Juan Pulhin. 'REDD policy impacts on indigenous property rights regimes on Palawan Island, the Philippines'. *Human Ecology* 40(5) (2012): 679–91.

18. Naomi Klein. *This Changes Everything: Capitalism vs. the Climate* (New York: Simon & Schuster Paperbacks, 2015), p. 203.

19. Ibid.

20. Laura Michéle et al. *When the SUN Casts a Shadow: The Human Rights Risks of Multistakeholder Partnerships: The Case of Scaling Up Nutrition (SUN)* (Heidelberg: FIAN International, 2019), www.fian.org/files/files/ WhenTheSunCastsAShadow_En.pdf (last accessed 16 April 2023).

21. Manahan and Kumar, *The Great Takeover*, 4.

22. Alem Tedeneke. 'World Economic Forum and UN sign strategic partnership framework', World Economic Forum, June 2019, www.weforum.org/ press/2019/06/world-economic-forum-and-un-sign-strategic-partnership-framework/#:~:text=The%20Strategic%20Partnership%20Framework%20 will,increase%20long%2Dterm%20SDG%20investments (last accessed 23 May 2023).

23. World Economic Forum. *Everybody's Business: Strengthening International Cooperation in a More Interdependent World* (Davos: Global Redesign Initiative, 2010), p. 9.

24. Mary Ann Manahan. 'Commodifying and selling nature to save it: multistakeholderism in global environmental governance', in *The Great Takeover*.

25. See the works of Prof. Ronald Mitchell of University of Oregon.

26. See 'The extractive industry transparency initiative in Papua New Guinea: just more corporate greenwashing?', ActNow, June 2015, https://actnowpng.org/ node/25597 (last accessed 27 May 2023).

27. For a good introduction about the Anthropocene, see Clive Hamilton, Christophe Bonneuil and François Gemenne (eds). *The Anthropocene and the Global Environmental Crisis: Rethinking Modernity in a New Epoch* (Abingdon: Routledge, 2015), http://site.ebrary.com/id/11055878.

28. The World Business Council on Sustainable Development is the largest CEO-led organisation of over 200 international companies, and connected to 60 regional and national corporate organisations. See 'Natural climate solutions', WBCSD, www.wbcsd.org/Programs/Climate-and-Energy/Climate/Natural-Climate-Solutions.

29. Ibid.

30. The three systems that require fundamental transformations include food, land and ocean use; infrastructure and the built environment; and energy and extractives.

31. See 'The future of nature and business', www3.weforum.org/docs/WEF_The_Future_Of_Nature_And_Business_2020.pdf.

32. See Heide Bachram. 'Climate fraud and carbon colonialism: the new trade in greenhouse gases', essay, December 2004, www.carbontradewatch.org/pubs/cns.pdf.

33. UNEP. 'Towards a green economy: pathways to sustainable development and poverty eradication – a synthesis for policy makers', 2011, www.unep.org/greeneconomy/Portals/88/documents/ger/GER_synthesis_en.pdf (last accessed 17 January 2023).

34. James Fairhead, Melissa Leach and Ian Scoones. 'Green grabbing: a new appropriation of nature?' *The Journal of Peasant Studies* 39(2) (2012): 237–61.

35. Bram Büscher and Robert Fletcher. 'Accumulation by conservation'. *New Political Economy* 20(2) (2015): 273–98.

36. Natural Capital Declaration. 'The Natural Capital Declaration', 2012, www.naturalcapitaldeclaration.org/wp-content/uploads/2013/12/The-Natural-Capital-Declaration-EN.pdf (last accessed 1 March 2023).

37. Robert Costanza et al. 'The value of the world's ecosystem services and natural capital'. *Nature* 387(6630) (1997): 253–60.

38. The protocol complements other national-level accounting frameworks such as the UN System of Environmental Economic Accounting (UNSEEA) implemented by governments through the World Bank-led Wealth Accounting and Valuation of Ecosystem Services (WAVES) global partnerships.

39. Peter Haas. 'Introduction: epistemic communities and international policy coordination'. *International Organization. Cambridge Journals* 46(1) (1992): 1–35.

40. Ibid. An example here is Swedish scientist Johan Rockström who co-produced the influential report *Planetary Boundaries and Global Commons in the Anthropocene*. It has been the basis of multistakeholder initiatives such as the Global Commons Alliance (GCA) that was created to transform the global economy, and maintain the resilience and stability of the planet's natural systems. It is governed by a leadership comprised of top business executives from the WEF and WBCSD, research and academic institutions such as the World Resources Institute (WRI), Potsdam Institute for Climate Impact

Research, Centre for Global Commons and international conservation giant, WWF International.

41. Gurminder Bhambra and Peter Newell. 'More than a metaphor: "climate colonialism" in perspective'. *Global Social Challenges Journal* (published online ahead of print 2022), https://doi.org/10.1332/EIEM6688.

42. Walter Rodney. *How Europe Underdeveloped Africa* (London: Bogle-L'Ouverture Publications, 1972).

43. See 'Roundtable on Sustainable Palm Oil (RSPO): 19 years is enough', CADTM, December 2022, www.cadtm.org/Roundtable-on-Sustainable-Palm-Oil-RSPO-19-years-is-enough (last accessed 27 May 2023).

44. Hariati Sinaga. 'Sustaining plantations and certifying inequalities: towards a decolonial critique of sustainable palm oil certifications in Indonesia.' Working Paper no. 17, Bioeconomy & Inequalities, (Jena, 2022), www.bioinequalities. unijena.de/sozbemedia/WorkingPaper17.pdf (last accessed 27 May 2023).

45. Manahan, 'Commodifying'.

46. June Rubis and Noah Theriault. 'Concealing protocols: conservation, Indigenous survivance, and the dilemmas of visibility.' *Social & Cultural Geography* 21(7) (2022): 962–84.

47. Getachew, *Worldmaking*, 188.

48. Through the pressures of social movements, trade unions and civil society organisations, in 2014, the UN Human Rights Council took steps to elaborate this legally binding instrument.

49. IPES-Food. 'Who's tipping the scales? The growing influence of corporations on the governance of food systems, and how to counter it', 2023, www.ipes-food. org/_img/upload/files/tippingthescales.pdf (last accessed 27 May 2023).

PART III

Horizons Toward a Dignified and Liveable Future

12

Resist Extractivism and Build a Just and Popular Energy Transition in Latin America

Tatiana Roa Avendaño and Pablo Bertinat

Latin America is a rich region in terms of experiences that have maintained or recovered the management of the common and proposed other forms of production and energy generation. There have been decades of debates and struggles against all kinds of projects that destroy ways of life, cultures, peoples and ecosystems. Communities have incorporated worldviews of indigenous and native peoples, who conceive oil as the blood of the earth and determine rivers as sacred, the care of territories, community management and agroecological production as being central to autonomy. These peoples and communities show us that the climate and environmental crisis can be taken on using their knowledge and wisdom, through different ways of relating to Nature, energy, water and food. All this implies rethinking what we have taken for granted until now, with the goal of promoting major cultural transformations and building futures where good living and a flavourful life are possible.

This chapter seeks to present several proposals that Latin American organisations and communities have raised and built around energy and energy transition. It begins with an account of the historical struggles against fossil fuels that have put the energy issue on the movements agenda, repositioning the understanding of energy itself, which is often considered a mere 'sector' of politics to be addressed from technical expertise, into the fabric of the reproduction of life and the interrelationships that sustain it. The chapter also discusses alternative forms of ownership and management, as well as measures required by states to make such alternatives viable and reach a fair, popular and sustainable energy model. Finally, it introduces the notion and experiences of community energies, which seek to recover control and management of energy from local social organisation.

PATHS TAKEN TO ABANDON FOSSIL FUELS

In the 2022 presidential campaign in Colombia, the world was surprised with the announcement by then candidate and now President Gustavo Petro that his first act of government would be to suspend new hydrocarbon exploration and ban fracking to move towards a de-escalation of the use of fossil fuels in the country. Although President Petro's approach amazed many people in Colombia and the world who considered it impossible to give up the oil path, the debates and proposals to leave oil in the ground and move towards post-oil societies date back several decades.

In June 1995, the U'wa people, who live on the border of Colombia and Venezuela, faced the possibility of experiencing oil production in their territory. In response, they issued a public manifesto in which they stated that they preferred 'a dignified death' rather than allow the exploitation of their land (Peoples Public Manifesto U'wa[1]). This manifesto was initially interpreted as a threat of collective suicide, but in reality, it was a declaration of their determination to fight against the Samoré oil block, which belonged to the Colombian state-owned oil company Ecopetrol and Occidental Petroleum Company (Oxy), a private US company. The oil block occupied 209,000 hectares in the eastern part of the Eastern Cordillera, in the department of Boyacá, and overlapped with U'wa ancestral territory and some of the areas where they reside.

The narrative behind the U'wa's resistance to oil was based on their worldview, which included a simple but forceful demand: they did not want the extraction of hydrocarbons in their territory, since extracting oil from the bowels of the earth would lead to the death of the Pacha Mama. For this indigenous people, oil is 'ruiría', which means 'the blood of the earth'. Their intense struggle, mobilisation and resistance put a stop to oil expansion in their territory. In Colombia, this indigenous struggle broadened the perspective of social struggles against oil exploitation, incorporating cultural elements such as the sacred, spiritual and symbolic value of the territory to put up resistance to oil production. This can be considered one of the first precedents of resistance to 'leave the oil underground'.[2]

In 1996, OilWatch, the South–South network of Resistance to Oil Activities was created in Ecuador, promoted by the organisations Acción Ecológica of Ecuador, and Earth Rights Action (ERA) of Nigeria. Both shared a radical position against extraction and the oil industry. Their fundamental criticisms were directed at the role that oil has played in modern capitalist development, the serious environmental liabilities that it causes in the territories and the planetary environmental crisis. In 1997, during the Conference of the Parties on Climate Change (COP) in Kyoto, Japan, OilWatch issued a statement calling for a moratorium on new oil projects. The proposal sounded radical and attractive to many environmental groups around the world, and hundreds of organisations signed the declaration. The moratorium declaration forced several of these groups to incorporate a central issue into the climate debate: to change the course of the unbridled oil burning in the world. Thus, the struggles against oil spread like seeds in the wind.[3]

In 2007, while a new political constitution was being discussed in Ecuador, indigenous, peasant and environmental organisations proposed the initiative to 'leave the oil underground' as a concrete solution for taking on the climate crisis. The proposal was based on the fact that industrialised countries, historically responsible for global climate change, should compensate non-industrialised countries holding oil reserves (such as Ecuador) to keep the oil underground, and the non-industrialised countries would use those financial resources to commit to transforming their productive matrix. Rafael Correa's administration formalised this proposal through the Yasuní Initiative, but did not have enough support from the European or North American governments to make real commitments. Meanwhile, although Ecuadorian ecological economists and environmentalists made proposals to gain support for this initiative, the Rafael Correa administration chose to continue with oil production in the Amazon and did not put enough effort into diplomacy consistent with the Yasuní Initiative.

Meanwhile, OilWatch presented documents questioning the oil-dependent society and explaining the urgency of leaving the oil underground to generate more sustainable ways of life. In 2015, at COP21 in Paris, OilWatch proposed creating an 'Annexo O' group[4] to recognise and respect the commitments and efforts of the communities and territorial organisations that are putting together projects that avoid extracting oil, gas or coal in efforts to avoid climate disaster.[5]

Taking this background into account, in 2022, the Colombian administration of Gustavo Petro Urrego and Francia Márquez Mina, from the Historical Pact, announced the suspension of new hydrocarbon exploration and promised to carry out a gradual, fair and orderly energy transition. During COP27, held in November 2022 in Sharm el-Sheikh, President Petro presented a decalogue of actions to face the climate crisis, where he highlighted the importance of abandoning the path of fossil fuels. Petro said that 'the climate crisis can only be overcome if we stop consuming hydrocarbons.

It is time to devalue the hydrocarbon economy with defined dates for its end and give value to the branches of the decarbonised economy. The solution is a world without oil and without coal.' He demanded that 'the world's private and multilateral banks (...) stop financing the hydrocarbon economy.'[6] His proposals once again bring the debate on 'leaving the oil underground' into the governmental sphere and even in the international negotiating agenda to address the climate crisis.

In this context, Censat Agua Viva, the Eco-social and Intercultural Pact of the South, and other organisations prepared the document 'Planned reduction of fossil dependency in Colombia: between cultural change and participatory demand management'. The text presents five proposals that lead in 'a process of no less than 15 years, to a socio-ecological transformation towards other forms of relationship between societies and all the species and living systems that make up our planet'.[7]

Undoubtedly, the struggles of the U'wa people and other local organisations in the world, as well as those organisations that are members of OilWatch, have led the way with a proposal that is gaining more and more strength: 'leave the oil underground' as a concrete measure to achieve sustainability and guarantee a true 'zero'.

WHAT IS REQUIRED TO CHANGE AND HOW?

From a socio-environmental justice perspective and within the universe of popular environmentalism, we defend a just and popular energy transition that is based on an anti-capitalist and socio-ecological narrative. However, to achieve this, we must first make a diagnosis of the current situation and establish the path towards a desired future. In this regard, it is important to understand the magnitude of the changes needed to address the problems associated with energy. This implies taking into account not only greenhouse gas emissions, but also social inequalities and socio-environmental impacts in the territories, as well as conflicts associated with energy and the concentration of energy power in a few hands and with large corporations.

We understand the energy system as a set of social relationships that bind us as a society and in our society–Nature relationships, which are determined by production relationships.[8] The just and popular energy transition requires decommodifying, democratising, defossilising, deconcentrating, decentralising and depatriarchalising. But what actions and processes are necessary to achieve that?

The path of decommodification and democratisation

The just and popular energy transition is based on the premise that all people have a right to energy, and challenges the idea that energy is a commodity. It is about deprivatising and strengthening the various forms of the public, the participatory and the democratic.[9] One of the slogans is to decommodify, which implies freeing energy from the predominance of the commercialised logic of economic benefit and focusing it instead on the ability to control and reproduce life in all its dimensions, both material and symbolic.[10]

We consider that energy is part of the commons, and therefore, it is a collective right in congruence with the rights of Nature. It is necessary to build a vision of energy as a right, taking the struggles for the right to water as an example. This right is not only for human beings, but for all living beings. We incorporate Nature and all its species into this definition, because we recognise that there is an interdependence between the full enjoyment of human life and the environment.[11]

Within the framework of the current capitalist system, markets are instruments that serve sectors whose rationale is based on unlimited capital accumulation, beyond physical limits and life. Therefore, capitalist markets are not neutral places.[12]

The concept of decommodification challenges the centrality of capitalist markets to solve certain needs. The recovery of the public is essential to this path. It not only implies a debate about ownership by reclaiming it from private hands, but also about management. In our perspective, recovering the public should not be limited to its association with the state (national). It is a question of strengthening and recreating all forms of the public, in terms of ownership and management, including historical experiences relating to the community, communal, municipal, collaborative and cooperative areas. These are valuable tools that must be strengthened in the face of the supposed superior efficiency that private companies offer in the provision of services.

Decommodifying and socially constructing the right to energy implies, among other tasks, a broad legislative, regulatory and normative reform that repeals privatisation laws and the liberalisation of markets that have placed the private sector at the centre of the energy system. It is also key to advance a de-privatisation process that includes not only energy companies but also other basic services, as well as developing tools that strengthen all forms of the public in terms of ownership and management, with emphasis on different levels and spheres (cooperative, community, state and national). It is necessary to strengthen the required institutional framework to achieve this.

As a first step towards a process of democratisation of the sector, it is necessary to establish information mechanisms that allow the participation of any community to be involved in decision-making, whether urban or rural. To do so, it is important to review, correct and even, on some occasions, reverse the direct subsidy policies for fossil fuels and various sectors of the fossil-based economy. It is also crucial to recognise and support institutions and actors involved in the generation, distribution, management and consumption of energy outside the capitalist market.

Furthermore, it is important to assume the possibility of deciding on energy at the local level, in its different dimensions (generation, consumption, energy poverty, etc.). Municipal energy agencies and experiences of reclaiming public services are examples that could be strengthened.[13] To make this process more dynamic, it is also necessary to advance methodologically: developing tools and procedures for constructing local, community and municipal energy policies as a form of collective appropriation of these policies.

It is not just about decarbonising

Carbon sinks, which are the mechanisms that absorb greenhouse gas emissions, and the finite availability of materials and minerals set a limit on the ability to substitute fossil fuels with renewable sources, within the framework of the current production and consumption matrix. This means that it is essential to reduce the net use of energy as the main goal, although this reduction must be planned and executed while taking into account the need to balance existing inequalities and the needs of different countries and social groups.

It is also important to take into account that it is not enough to merely advance in the use of renewable energy sources. Rather, it is necessary to consider the environmental, social and political dimensions of each specific venture to determine its sustainability.

Among the actions that can be taken to face these challenges, the following stand out:

(i) agree not to exploit unconventional and conventional hydrocarbons in risk-prone areas, such as offshore zones, or reduce their use within the framework of a plan to abandon fossil fuels in the short term;

(ii) monitor the net decrease in energy use beyond the climate commitments made;

(iii) have specific proposals for different sectors, such as transportation, which in Latin America is the main energy consumer and should be considered as an energy sector in and of itself;

(iv) develop tools that visualise the socio-economic benefits of energy efficiency and establish regulatory changes that go against commercial logic;

(v) stop adopting renewable energy competitive bidding processes between large commercial/transnational providers as the only option and prioritise instead the decentralised and deconcentrated development of these sources.

On the production model and consumption

In order to move towards a just and popular energy transition, it is necessary to build a production model that is compatible with the sustainability of life and the care of the ecological systems and cycles that make it possible. It is essential, as feminists propose, to *put life at the centre* of this model.

The energy transition that we propose requires recognising the 'natural and human physical limits, as well as the immanence and importance of links and relationships as inherent features of the existence of life'.[14] These conceptions are associated with new ways of organising life in society, new forms of production, revaluation of the place occupied by productive and reproductive work in societies, and new forms of consumption, associated with a change in the society–Nature metabolism.

Regardless of the initiatives associated with energy efficiency in various sectors, it is necessary to advance in sectoral analyses to question the regional production and transportation matrix and seek sustainable and fair alternatives. Concrete proposals in this area include, for example:

(i) establish maximum circuits for the circulation of goods and develop short production chains that prioritise local products;

(ii) analyse the areas of material production that need to degrow and determine what to stop producing; analyse how to enhance services over material goods. This must be accompanied by establishing timelines for this degrowth (see the chapters by Bengi Akbulut and Luis González Reyes in this book);

(iii) develop new areas of production and less energy-intensive services;

(iv) establish timelines to stop using individual internal combustion vehicles;
(v) implement a process of modal change in freight transport;
(vi) rethink the role and design of infrastructure, since it is financed with public funds and determines future behaviour and consumption.

Similarly, a process must be undertaken that allows us to advance in the social construction of other forms of satisfying human needs. It is an intense and extensive process, but one that can be streamlined through the use of various tools, for example, strengthening urban networks for sustainable consumption; developing regulations that prohibit planned obsolescence; making mass life cycle analyses of products; prohibiting or restricting advertising on particular branches of products; establishing a rapid program to eliminate energy poverty; associating energy policies with housing policies; and restricting luxury uses of energy.

THE PATH PROPOSED BY COMMUNITIES TO ADDRESS ENERGY

In Latin America, various social, ethnic and community organisations, both urban and rural, have implemented local initiatives for sustainable, autonomous and decentralised energy. These experiences address the relationship with energy in conjunction with other practices and processes of life reproduction, such as the transformation of productive and agri-food systems and care for Nature, through organised community processes that promote the generation of energy in an autonomous and decentralised fashion.[15]

But how is community-based energy built in practical terms? There are many possibilities, which include the promotion of non-conventional renewable energy technologies that include photovoltaic systems, small hydroelectric power plants and wind turbines, as well as various technologies adapted to local conditions, such as biodigesters, solar dryers or dehydrators, solar water heaters, pedal-powered machines, and techniques associated with energy forest plantations/wood fuel forests, agroecological production, digesting bales, mobility and sustainable architecture and water harvesting and management.

Community-based energy sources are based on collaboration, mutual care and solidarity. Juan Pablo Soler[16] refers to them as 'Ubuntu Energetics', because they exist to the extent that the other exists. Such energy sources seek to regain autonomy, challenging the current energy system, which is centralised, concentrated, undemocratic, racist/colonialist, patriarchal and authoritarian.

The energy transition proposed by the processes that drive community-based energy requires a cultural transformation in terms of the generation, use and the very concept of energy. This transition also implies the democratisation and participation around energy, the relocation of activities and local control. According to Sandra Rátiva,[17] it is a social and collective reappropriation of energy that consists of producing and rescuing social and ecosystem links that promote the development of local economies, knowledge production, coexistence and improvement of living conditions. In other words, it seeks to 'reconnect the ecological, the social and the political'[18] to face multiple crises: climate, energy and environmental.

Community-based energies ultimately propose a perspective of the world that has a greater understanding of the physical limits that we share with the rest of the living world. They recognise the urgency of stopping the unbridled rate of growth, as well as establishing more harmonious relationships with Nature. In addition, they propose that local communities must become active subjects of the energy system, ceasing to be mere consumers of energy, assuming a role as producers of their own energy. In this perspective, community energies invite us to rethink energy and associate other dimensions to it that go beyond electricity.

CONCLUSION

The just and popular energy transition is an urgent necessity to address the climate and environmental crises we face. To be successful in this transition, a productive model compatible with the sustainability of life and the care of ecological systems is needed, placing life at the centre. It is crucial to build different ways of life and relationships around energy, and to recognise that this transition is not only a technical process, but also a social and cultural one, which implies the construction of other ways of understanding and satisfying human needs.

To achieve a just and popular energy transition, a series of actions and processes must be carried out, from the states and organised society, that promote the decommodification, democratisation, defossilisation, deconcentration, decentralisation and depatriarchalisation of energy. It is essential to build a vision of energy as a collective right consistent with the rights of Nature and recover the public in terms of ownership and management, including historical experiences that are community, communal, municipal, collaborative and cooperative.

To achieve this, states, legislative, regulatory and normative reforms are needed: actions that repeal the laws of privatisation and liberalisation of the markets that have given rise to the dominance of the private sector in the energy system. It is also necessary to review and rectify the policies of direct subsidies to fossil fuels and establish information mechanisms that allow public participation in decision-making. It is equally important to recognise and empower institutions and actors outside the capitalist market and develop construction tools and processes for a just and popular energy transition.

The debates and proposals towards energy and socio-ecological transitions from the people are making progress and disputing perspectives and narratives with other logics of understanding transitions. They also constitute an invitation to work together to construct initiatives that ensure a sustainable future for all people and the planet.

NOTES

1. Margarita Serje. 'ONGs, indios y petróleo: el caso U'wa a través de los mapas del territorio en disputa'. *Bulletin de l'Institut Français d'Études Andines* 32 (2003), https://doi.org/10.4000/bifea.6398.

2. Tatiana Roa Avendaño. 'Soberanía y autonomía energética. Treinta años de debates alrededor de asuntos cruciales', in Tatiana Roa Avendaño (ed.), *Energías para la transición. Reflexiones y relatos* (Censat Agua Viva, Fundación Heinrich Boll, Oxfam, 2021).

3. Ibid.

4. The United Nations Framework Convention on Climate Change, signed at the Rio Summit in 1992, divided the states parties into two groups: Annex I and Non-Annex I. Annex I includes industrialised countries from the North and some in transition, while Non-Annex I covers mainly countries from the Global South. Later, Annex II was established, which includes Annex I members, to provide financial and technical resources to countries of the South to reduce greenhouse gas emissions. According to OilWatch, the engine of the development of capitalism for Annex I and II countries has been the exploitation of fossil fuels, and they have done everything possible to stop any measures that limit their consumption. Despite decades of international negotiations, global warming continues to increase without effective solutions to stop it. OilWatch's Annex 0 Group proposal would give a voice in the negotiations to the peoples and communities that are solving the causes of the climate crisis. See: www.oilwatch.org/es/2015/07/16/es-tiempo-de-crear-el-grupo-anexo-o/.

5. Oilwatch. 'Propuesta para la COP21 de París, diciembre 2015. Es tiempo de crear el grupo "Anexo Cero"', 2015, www.oilwatchsudamerica.org/images/stories/ANEXO_CERO_OILWATCH_Espanol.pdf.

6. Gustavo Petro. 'Discurso ante la Conferencia de las Partes de Cambio Climático en Sharm el Sheik', Cancillería.gov, 8 November 2022, www.cancilleria.gov.co/newsroom/news/presidente-petro-aseguro-cop-27-egipto-enfrentar-crisis-climatica-solucion-mundo.

7. Censat Agua Viva, Pacto Ecosocial del Sur, Energía y Equidad, OilWatch. 'Disminución planeada de la dependencia fósil en Colombia: entre el cambio cultural y la gestión participativa de la demanda', Censat Agua Viva, September 2022, https://censat.org/disminucion-planeada-de-la-dependencia-fosil-en-colombia-entre-el-cambio-cultural-y-la-gestion-participativa-de-la-demanda/.

8. Maristella Svampa and Pablo Bertinat. *La transición energética en Argentina* (Buenos Aires: Siglo Veintiuno Editores, 2022).

9. Ibid.

10. Miriam Lang and Raphael Hoetmer. 'Buscando alternativas más allá del desarrollo', in Mirian Lang and Raphael Hoetmer (eds), *Alternativas en un mundo de crisis* (FRL: Universidad Andina Simón Bolívar, 2019), p. 927.

11. Pablo Bertinat, Jorge Chemes and L. Moya. 'Derecho y Energía'. Curso de formación en energía, 2012, unpublished manuscript.

12. G. Aguirrezábal and S. Arelovich. 'Desmercantilización. Aproximaciones al estado del debate. El caso particular del sector energético', Taññer Ecologista, 2011, unpublished manuscript.

13. See Satoko Kishimoto and Olivier Petitjean. 'Reclaiming public services. How cities and citizens are turning back privatisation'. TNI, 2017, www.tni.org/en/publication/reclaiming-public-services.

14. Yayo Herrero-Lopez, 'Prólogo. Repensar la vida en tiempos de emergencias', in Tatiana Roa (ed.), *Energías para la transición. Reflexiones y relatos Avendaño* (Censat Agua Viva, Fundación Heinrich Boll, Oxfam, 2021), p. 12.

15. Information about several of these experiences is available in the 'Virtual Exhibition of Just Energy Transition (TEJ) Community Experiences', promoted since 2020 by Censat Agua Viva, the Energy and Equity Group, the Foundation for Intercultural, Educational and Environmental Expression, Fundaexpresión, Comunidades Sembradoras de Territorio, Agua y Autonomía, Setaa and the Movement of People Affected by Dams in Latin America, MAR (see https://transicionenergeticajusta.org/). These experiences contribute to enriching the debates and perspectives on the energy and socio-ecological transitions, questioning the proposal of a transition limited to the change of the energy matrix based on the source of energy and propose to address consumption patterns, end the centralisation of energy and promote its democratisation. These community experiences also propose new narratives, tools and methodologies to achieve the energy transition and energy sovereignty.

16. Juan Pablo Soler. 'Ubuntus energéticos en audiencia pública'. Revista Raya, 8 November 2022, https://revistaraya.com/juan-pablo-soler/155-ubuntus-energeticos-en-audiencia-publica.html

17. Sandra Rativa-Gaona, 'La interdependencia como una clave para pensar la transición energética', in *Energías para la transición. Reflexiones y relatos*, p. 181.

18. Ibid.

13

Eco-feminist Perspectives from Africa

Zo Randriamaro

This chapter unpacks the African debates on eco-feminism and looks at its contemporary concepts and practices which are rooted in the African context. It explores the ways in which they are shaping transformative pathways towards a new Pan-African decolonising movement from below that fully embraces Afro-feminist politics, and in this sense, frame multi-dimensional alternatives to hegemonic and unjust transitions. The chapter also shows how this eco-feminist movement builds on an African philosophy such as Ubuntu to offer a living alternative and a different future, centred upon collective solidarity and sharing between peoples, together with truly sustainable modes of living in harmony with Nature. The first part of the chapter explores the roots, current concepts and practices of African eco-feminisms, including their struggle against the new wave of green extractivism; the second part discusses some of the major African ecofeminist strands of thought and political agendas. The last part focuses on the African worldviews and values enshrined in the Ubuntu philosophy, and how they underpin contemporary African eco-feminist movements struggling for just, equitable and sustainable alternatives to the dominant neoliberal and extractivist development system.

AFRO-FEMINISMS AND ECOLOGY

For a better understanding of African eco-feminisms, it is important to start with the recognition of their ancient and Pan-African filiations that are often forgotten and subsequently erased from the history and collective memory of the global eco-feminist movement. It is equally important to explore their current concepts and practices, which are anchored in their respective political, social and economic contexts, and how they relate to ecological debates and movements.

The roots of African eco-feminism

Of central importance to African eco-feminism is 'the relational living between and among human beings and other physical and animate aspects of the environment',[1] especially between women and Nature. Although in the Eurocentric historical framework, eco-feminist theory is associated with the 1970s, the association between women and Nature has been made long before that period by feminist movements in Eastern Africa. These include political and religious struggles such as Ethiopianism, Nyabingi, the Mau Mau movement and New Rastafari, influential resistance movements during the modern colonial era, even as they had been hindered by slavery and European colonial interests.[2] I will shortly introduce them in the following paragraphs.

The late 1800s saw the rise of a religious movement called Ethiopianism resisting colonialism, within which the anti-colonial activists were organising around the idea of self-rule. It encompassed not only Ethiopia and the African continent, but also people of African descent in the Caribbean. In Jamaica, for example, women-led resistance movements embodied the dimension of gender in class struggles, such as in the Culture of the Free Villages,[3] which claimed the defence of the land by a majority of women.[4]

These movements took inspiration from the Nyabingi women in Uganda who unified against colonial oppression between 1910 and 1930, especially those who were involved in food production in rural areas and were the most vulnerable to colonial rule. Later in Kenya, rural women also organised around land ownership issues, while other women joined the Mau Mau movement, a militant nationalist movement against British domination in the 1950s, to avoid being traded as wives by their fathers.[5]

Thus, ideas of eco-feminism are visible in the link between women, land, food production and political rights that was made a long time ago in the Eastern African and Jamaican contexts by different African and Caribbean rural women's movements, mobilised in the struggle for their land rights and against colonial oppression.[6]

African eco-feminism poses the following question to feminists: how do we remove our power and energy from the dominant male economic and power development model, which has historically evolved towards an accelerated destruction of the very basis of life on the planet? At the same time, African eco-feminists also challenge certain ecologists, questioning the theoretical vision of an abstract 'human' who is dominant and oppressive over Nature and inviting to consider instead the multiple oppressions within human society itself, starting with the gender system which is a root cause of domination and oppression.

As such, African eco-feminisms are part of political movements and initiatives that are committed to the deconstruction of relations of domination at the intersection of gender, class, ethnic origin, and 'races', including colonialism, North/South hierarchies, etc., with the aim of overcoming all kinds of oppression against women and Nature.

Current concepts and practices

African eco-feminisms integrate a critical reflection on the links between the dominant 'development' model, the ecological crisis and the stakes of peace and non-violence, which allows them to come up with innovative and radical questionings, both on what feminism is and on how to approach ecology and Nature, or, as discussed in this book, processes of eco-social transition.

While the broader global movement is sometimes distracted by a divisive debate about whether gendered associations with Nature are reductive of women, it appears that most of the movements engaged in feminist and environmental activism in Africa have simply sought to create strategic and political alliances between women, Nature and environmental protection.

The Kenyan Wangari Maathai (1940–2011) and her Green Belt Movement (GBM) arguably represent the collective and ecocentric activism that defines the essence of African eco-feminism. As the first environmentalist to win the Nobel Peace Prize in 2004 for her contribution to sustainable development, democracy and peace, Wangari Maathai has drawn attention to the close relationship between African feminism and African environmental activism, which challenges both patriarchy and neo-colonialist structures that undermine the continent. As the South African gender specialist Janet Muthuki puts it, 'Wangari Maathai's GBM is an African ecofeminist activism, which through environmental issues and interventions highlights gender relations and challenges patriarchy within national and global ideological structures.'[7]

Maathai is also known for weaving traditional beliefs on nudity and gender[8] together with contemporary political struggles to foment a decisive moment in the struggle that brought women into the centre of the political arena in which previously they had only been marginalised. By so doing, she ultimately made a critical contribution to promoting the democratic movement in Kenya.[9]

Intersectional eco-feminism emphasises the importance of gender, race and class, and draws a strong connection between feminist concerns, human oppression within patriarchy, and the exploitation of the natural environment which women are more often dependent on – but of which in many cultural settings they are also considered the guardians. Because women experience the multiple crises facing Africa intersectionally, an intersectional approach to building radical movements for change is crucial. As stated by one of the intersectional eco-feminism leaders:

> What we need are truly radical and revolutionary transnational movements, not small cocoons. Of course, it's important to pay attention to local realities. In a very narrow scope, an eco-feminist movement to me is concerned with transforming the ways in which economic, intellectual and ecological resources are accessed by women, especially those most vulnerable and often on the frontlines of ecological devastation and climate change. It also means constantly working to re-claim and re-imagine much more just and egalitarian ways of being with one another and fundamentally, for me, that means destroying patriarchy and reclaiming ideas of the commons.[10]

The anti-extractivist dimension is another element of the conceptual framework that characterises the struggles of contemporary African eco-feminist movements, which is central to the debates around global justice and eco-social transitions. It is embodied in the work and political position of the African WoMin Alliance (WoMin) described below:

> Africa and other parts of the Global South are subject to another round of deepening colonisation as corporations and their host governments in the Global North and parts of the Global South chase the untapped and highly profitable frontiers of Africa's mineral and natural wealth. This gives impetus to what WoMin calls an extractivist development model, just another link in the chain of perpetual colonisation and exploitation of Africa and its peoples... Extractivism is deeply patriarchal and racist, relying on the cheap paid labour of Black and brown working-class men who work under extremely exploitative and dangerous conditions to guarantee profits for large multinational corporations and their vast supply chains. The unpaid care work of women is also incorporated into the accumulation of profits to corporations and the rich as they labour to

reproduce workers and their families through subsistence food production, the provisioning of water and fuel, and their care work...[11]

In the context of climate change and the resulting transition to renewable sources of energy, African eco-feminist activists are increasingly involved in struggles against mega-extractivist projects for the production of 'clean/green energy' like solar, wind, thermal and hydrogen. A case in point is the extraction of rare earths, which are in high demand for the energy transition in the Global North, particularly for the production of critical components of green technologies such as wind turbines, solar panels and hybrid fuel-cell batteries. The tremendous harms generated by contemporary rare earth production practices include massive land grabbing, pollution and destruction of ecosystems, and the resulting loss of livelihoods, in addition to the devastating effects on the health of people living nearby and downstream of rare earth mining and processing operations, such as cancers, birth defects and the decomposition of living people's musculoskeletal systems.[12] Against this background, eco-feminist activists are supporting the resistance of the affected communities against a mega-project for rare earths exploitation in Madagascar, where the project mining site is most likely to become a sacrifice zone to pay for the social, economic and ecological costs of rare earths mining for the Global North's consumption.

BUILDING A NEW PAN-AFRICAN DECOLONISING MOVEMENT FROM BELOW

The African eco-feminist movement is located at the confluence of three distinct movements that are fighting against the same imperialist ideologies and institutions that disrupted and undermined indigenous cultures and institutions: the anti-neoliberal movement, mainly supported by climate justice activists, the anti-imperialist movement brought forward by decolonialists and the anti-patriarchal movement protagonised by feminists. As such, Afro-eco-feminists are striving to dismantle power structures and hierarchies that oppress and exploit both women and Nature.[13]

A Pan-African feminist movement for climate justice

At the community level, there is a growing awareness of the threats to biodiversity and climate resilience resulting from large-scale, agro-industrial and extractive projects across the African continent, and their links with corporate and state powers. Eco-feminism is inseparable from the concrete struggles and initiatives at the grassroots to preserve, develop or repair liveable spaces and social bonds through material and cultural processes that allow a society to reproduce itself without destroying other societies or living species.

From this point of view, special attention should be given to climate justice movements which focus on the ecological crisis and its root causes, from a feminist perspective, based on the growing awareness among the affected people that the dominant neoliberal development model is unsustainable. Such eco-feminist movements centre on the climate and ecological crises in Africa, on their links to extractivist development and its gendered impacts, and demand 'that the unjust capitalist system be dismantled in order to take care of the planet and provide redress for historical violations of the rights of peoples and nature'.[14]

Because of their transnational nature, both the climate justice movement and the decolonisation project for Africa cannot be limited to a piecemeal approach but require a Pan-African course of action. The fragmentation and ideological divisions of the continent have greatly contributed to perpetuate the different forms of colonialism in Africa, which implies that Pan-Africanism is a critical pillar of the decolonisation project embraced by Afro-eco-feminists.

African eco-feminisms and decolonisation[15]

Wangari Maathai affirmed that 'colonialism was the beginning of the deterioration of nature due to industrialisation and the extraction of natural resources... Logging of forests, plantations of imported trees which destroyed the eco-system, hunting wildlife, and commercial agriculture were colonial activities that destroyed the environment in Africa.'[16] Thus, from the outset, Afro-ecofeminism has been an important pillar of a decolonial[17] feminist approach to promoting systemic change in Africa.

In this regard, Afro-eco-feminists have also been relying on their rich traditional heritage and indigenous culture to challenge patriarchal power and neocolonialism. While some African feminists have argued that the African cultural tradition and communitarian philosophy are not compatible with feminism because they are deeply patriarchal,[18] other eco-feminists affirm that African traditional philosophies and tools such as Ubuntu can be used to achieve gender justice as well as the other goals of Afro-feminism.[19]

As the Ugandan academic and Human Rights activist Sylvia Tamale argues, 'the underlying features of ecofeminism very much resembled those traditionally practised in non-Western Indigenous cultures'. In particular, eco-feminist practices have a lot to draw from 'the epistemic relationship between Indigenous people and nature (that) manifests through their spirituality, clan totems, taboos, ancestral myths, rituals, fables, and so forth ... Notably, the consequences of violating a taboo were not individualized and responsibility to conform was communalist. If you transgressed social taboos, your relatives would also suffer the consequences...'[20]

A typical illustration of this epistemic relationship is the statement below, expressed by the women who are the guardians of the local sacred sites and bio-cultural heritage (*Mpijoro tany*) of the indigenous group from the Sakatia island in the northwestern part of Madagascar:

> Our role as 'Mpijoro tany' is our duty to our village, which has been founded by our forefather. There is a sacred place called Ankatafabe, and there is another one in Ampijoroa, and also in Ankofiamena. In the past, there was no church but these were the places where we prayed to God, just like we do in a church. These are the places of annual 'fijoro-ana' (ritual prayer ceremony) to pray and to request benedictions... Our ancestors strictly observed 'fadin-tany' (land taboos), and most people in Sakatia still observe them. If a person breaks a 'fady' (taboo), he must kill a zebu in reparation of the wrong he has done. (Justine Hamba, ritual prayer leader – Sakatia)[21]

The other guardian of the sacred sites on Sakatia island explained as follows the rationale behind the traditional rituals and customs, and the vital importance of abiding by them for the common good and ensuring unity, cooperation, love and trust in the community, as well as in order to establish respect between the living and the dead:

> there is a way of preserving 'kodry' (fish) for people who eat them. You pick only the quantity you need; any surplus must be distributed to the community; it cannot be thrown away or sold. This is the sense of community and love. Those who pick the food are not necessarily the ones who eat it; it must be shared with the community. It cannot be sold and it cannot be harvested in large quantities; otherwise it will become extinct and by doing so, people do harm to the environment... The small animals in the village cannot be killed without any reason, for example

the 'Anjava' which is a small animal that lives in shady and cool places. The green forest where it hides should not be cut down. If a person kills such animal, then something bad will happen to him/her. The curse will not go away unless he/she takes away the punishment (manala fady) and apologizes to the traditional prayer leaders in the village... The person who broke the taboo commits a desecration; these are treasures of this land that our ancestors cherished and these animals should always be respected and remain in the village... It is forbidden to destroy forests that provide rain and fresh air we need for living. That's why Sakatia is a green island, because we don't cut down forests over the hills, and we also plant trees. And we also protect marine life, including fish, we prevent fishermen from using non-standard nets from getting here. We protect sea turtles, and endemic fish species like 'Horoko' and 'kodry'... We have a dina (traditional social convention with a system of sanctions) in the village: for example, if you swear or use foul language, there is a corresponding penalty in the 'dina'. You must go to the ritual prayer leaders and ask for apology, otherwise everyone in the village will be under curse. (Célestine, ritual prayer leader – Sakatia)[22]

As evidenced by the above statements, the Malagasy communities in Sakatia are abiding by the same 'ethics of nature-relatedness'[23] as the numerous indigenous groups in sub-Saharan Africa that are also wary of anthropocentric interventions on Nature which undermine the healthy web of life in ways that threaten the survival of the planet. As Sylvia Tamale has rightly underlined, 'women in the global South may not have self-identified as "ecofeminists", but they have a long history of ecological consciousness and moral obligation towards future generations.'[24]

AFRICAN ECO-FEMINIST ALTERNATIVES TO DEVELOPMENT[25]

From a decolonial, eco-feminist perspective, many rich alternatives already exist at the micro- and meso-levels. Many of these alternatives were taken from Africa, such as the solidarity economy and collective solutions to labour and resources like seeds and money, and must be recognised and built on. As happened in Latin America with other proposals adopting some of the positions and cosmovisions of indigenous peoples, including the rights of Nature and the worldview of 'Buen Vivir' (a Spanish phrase that refers to a good life based on a social and ecological expanded vision),[26] there is certainly a significant African archive of endogenous ideas, practices and political concepts that lie in tradition, as well as in anti-colonial struggles and post-colonial transformations from which we should draw inspiration and guidance. These include indigenous knowledge systems, communal tenure/indigenous land rights and social labour cooperation.

Chief among these are the critical alternatives based on 'the African worldview and philosophy known as Ubuntu in Southern Africa'[27] which is largely practised across sub-Saharan Africa[28] and 'tries as much as possible to whittle down traditional patriarchal, dualist and anthropocentric views of existence'.[29] Owing to Ubuntu, Africans have celebrated the values which connect past and present as well as humans and Nature for centuries.[30]

As an African ethical paradigm, Ubuntu is not compatible with capitalist relations,[31] private property[32] and pervasive inequality.[33] Rather, it demands an activism of solidarity and decolonisation in the face of what Vishwas Satgar terms an 'imperial ecocide'.[34] Ubuntu's ecological ethics has generated 'the radical notion of post-extractivism, that is, leaving behind for future generations the fossil fuels and minerals that drive destructive capitalist accumulation and its crises, notably climate change'.[35]

From an eco-feminist perspective,

> Ubuntu environmental ethics seek to emphasise the need to treat various aspects of nature that have traditionally been considered as morally insignificant – such as non-human animate beings – with care, reverence, kindness and accord them ethical consideration. At the same time, this ecofeminist dimension in ubuntu implies that similar values that emanate from the virtues of ubuntu – such as caring, goodness and reverence – could also be accorded or ascribed to non-animate aspects of nature such as physical nature, plants and water bodies that do not necessarily have sentience.[36]

In particular, living alternatives are already proposed by African rural and indigenous women in the defence of their territories, their autonomy, their forms of production, their community relations and their interdependent relationship with Nature without which they would not survive, against the deeply destructive extractivist model. Such living alternatives can be identified in the ways in which they produce, exchange, care for and regenerate our natural resources; nurture our families and communities; cooperate in our communities, etc. As WoMin puts it, 'the majority of women in Africa, who carry the burden of the climate and ecological crisis and who have paradoxically contributed the least to the problem, are practicing and proposing, in their deeply ecofeminist resistance to extractivist patriarchy, a development alternative which all humanity must respect and echo if we and the planet are to survive.[37]

In concrete terms, just and sustainable alternatives for a different future, which would be built on the philosophy of Ubuntu and centred upon a collective solidarity and sharing between peoples, together with truly sustainable ways of living in harmony with Nature, would include a series of elements proposed by African eco-feminists.[38] First of all, they would enforce food sovereignty, through an agro-ecological low-input model of agriculture. They would guarantee people's sovereignty over their own path towards well-being, through the concept of consent for women in the Global South, which gives credence and space to lived development alternatives at the local level. At the same time, those alternatives would have to aim at energy sovereignty through sustainable and decentralised collective forms of generation of renewable energy under the control of communities and specifically women, and put an end to the extraction and burning of all fossil fuels. They would still allow small-scale, low impact forms of extraction, under collective forms of ownership and subject to local and regional priorities. In terms of their governance model, they would have to put forward participatory, inclusive democracy at all levels of decision-making, which recognises women's central role in society, their different needs and the requirement for full and ongoing consent by affected communities and women in particular.

Those alternatives would also challenge the primacy of private property, respecting and supporting systems in which natural resources are 'owned' and managed by collectives and groups, and the active expansion of common properties as a critical part of the fight against privatisation and financialisation. And they would promote and enforce degrowth and a rapid transition to a low consumption lifestyle on the part of the rich and middle classes in the traditional Global North and South.

NOTES

1. Munamato Chemhuru. 'Interpreting Ecofeminist Environmentalism in African Communitarian Philosophy and *Ubuntu*: An Alternative to Anthropocentrism'. *Philosophical Papers* 48(2) (September 2018).

2. Amélie Gontharet. 'An ecofeminist perspective of the impact of development policies on women's lives. The case of Ethiopia', Master's Dissertation, University of Cyprus, European Master's Degree in Human Rights and Democratisation, 2018, 16.

3. Inspired by Mrs James Mckenzie in 1902 and the Nyabingi women in Uganda among others.

4. Gontharet, 'An ecofeminist', 17.

5. Ibid.

6. Ibid.

7. Janet Muthoni Muthuki, 'Rethinking ecofeminism: Wangari Maathai and the Green Belt Movement in Kenya', Master's Dissertation, University of Kwazulu-Natal, 2006.

8. Ibid.

9. Nyabola. 'Wangari Maathai was not a good woman. Kenya needs more of them... she revolutionised the act of protest in Kenya by centring it on the female body. In urging the protesting mothers of detainees to strip when threatened by security officers who were threatening to break up their protests...'

10. Jessica Merino. 'Women speak: "Ruth Nyambura insists on a feminist political ecology"'. MS Magazine, 15 November 2017, https://msmagazine.com/2017/11/15/women-speak-ruth-nyambura-feminist-political-ecology/.

11. Margaret Mapondera, Trusha Reddy and Samantha Hargreaves. 'If another world is possible, who is doing the imagining? Building an ecofeminist development alternative in a time of deep systemic crisis'. Bread & Butter Series: African Feminist Reflections on Future Economies, Accra, 2020, 2.

12. Julie Michelle Klinger. Rare Earth Frontiers. From Terrestrial Subsoils to Lunar Landscapes (Ithaca, NY: Cornell University Press, 2017).

13. Sylvia Tamale, Decolonization and Afro-Feminism (Ottawa: Daraja Press, 2020).

14. Mapondera et al., 'Ecofeminist', 9.

15. The term 'decolonisation' used in this chapter is defined by Sylvia Tamale as 'a multi-pronged process of liberation from political, economic and cultural colonization. Removing the anchors of colonialism from the physical, ecological and mental processes of a nation and its people' (Tamale, Decolonization, 14).

16. Wangari Maathai. The Challenge for Africa (New York: Pantheon Books, 2009), pp. 68–69.

17. According to Sylvia Tamale, 'Decoloniality: A specific type of decolonization which advocates for the disruption of legacies of racial, gender and geopolitical inequalities and domination' (Tamale, Decolonization, 14).

18. Fainos Mangena. 'The search for an African feminist ethic: a Zimbabwean perspective'. Journal of International Women's Studies 11(2) (2009): 18–30.

19. Such as Sylvia Tamale and Munamato Chemhuru.

20. Tamale, Decolonization, 87–89.

21. CRAAD-OI (Centre de Recherches et d'Appui pour les Alternatives de Développement – Océan Indien or The Center for Research and Support for Development Alternatives – Indian Ocean). 'Women's dialogues to dream and imagine development alternatives'. Report of the rollout process in Sakatia, Madagascar, 2021.

22. Ibid.

23. Segun Ogungbemi. 'An African perspective on the environmental crisis', in Louis Pojman (ed.), Environmental Ethics: Readings in Theory and Application (Belmont, CA: Wadsworth Publishing Co., 1997), pp. 330–37.

24. Tamale, *Decolonization*, 85.
25. This section draws heavily from the Concept Note on Development Alternatives developed in May 2020 by Zo Randriamaro for WoMin.
26. Eduardo Gudynas. 'Debates on development and its alternatives in Latin America: a brief heterodox guide', in Miriam Lang and Dunia Mokrani (eds), *Beyond Development. Alternative visions from Latin America.* (Quito and Amsterdam: Rosa Luxemburg Foundation and Transnational Institute, 2013), p. 35.
27. Christelle Terreblanche. 'Ubuntu and the struggle for an African eco-socialist alternative', in Vishwas Satgar (ed.), *The Climate Crisis: South African and Global Democratic Eco-sociaist Alternatives* (Johannesburg: Wits University Press, 2018), p. 168.
28. Leonard Tumaini Chuwa. *African Indigenous Ethics in Goba Bioethics: Interpreting Ubuntu* (New York: Springer, 2014).
29. Chemhuru, 'Ecofeminist', 19–20.
30. Tamale, *Decolonization*.
31. Laurence Caromba. 'Review of Ubuntu: curating the archive'. *Strategic Review for Southern Africa* 37(1) (2014): 208–11.
32. Dorine Eva Van Norren. 'The nexus between Ubuntu and global public goods: its relevance for the post-2015 agenda'. *Development Studies Research* 1(1) (2014): 255–66.
33. Drucilla Cornell and Karin Van Marle. 'Ubuntu feminism: tentative reflections'. *Verbum and Ecclesia* 36(2) (2015).
34. Satgar, 'Climate crisis'.
35. Terreblanche, 'Ubuntu', 169.
36. Chemhuru, 'Ecofeminist', 20.
37. Mapondera et al., 'Ecofeminist'.
38. WoMin. 'Women Building Power: exploring our initial ideas about an African ecofeminist approach to campaigning', discussion document, n.d.

14

A Feminist Degrowth for Unsettling Transition

Bengi Akbulut

INTRODUCTION

Transition is indeed the buzzword of our time. While far from being uncontested, the term has also increasingly and visibly been appropriated by corporations, nation-states and international organisations of the *status quo*. Transition is now being invoked in ways that risk perpetuating global environmental injustices and neocolonial dynamics of resource appropriation, and open novel fields of capital accumulation while shifting the socio-ecological burden of transition to the Global South (and/or the South within the North). The danger that lies with such uses and circulation of the term, however, is not limited to the practices justified by the vaguer and tamer message it carries. The hegemonic buzzing of transition, so to speak, also crowds out historical and contemporary worldviews, struggles and proposals that emerged, flourished and have been practised, both in the Global South and North. One such proposal is degrowth.

'The global equilibrium', wrote André Gorz in 1972, 'for which no-growth – or even degrowth – of material production is a necessary condition, is it compatible with the survival of the (capitalist) system?'[1] Since this first use of the term, 'degrowth' – or in its French original *décroissance* – has become a forceful conceptual framework and a political mobiliser for imagining and enacting alternative ways of articulating society, economy and Nature. The notion has since entered academic literature, vocabularies of social movements and public debate (even in the European Parliament). The academic literature on degrowth, in particular, has reached an impressive volume and scope, ranging from issues of infrastructural adjustment and reorganisation of work to the design of monetary systems and a new architecture of public finance.

This chapter situates degrowth as a counterhegemonic proposal that unsettles and goes beyond dominant understandings of transition. Emphasising an understanding of degrowth as one of recentring and reorienting the economy (rather than merely a matter of biophysical downscaling), the chapter delineates three axes that are fundamental for this potential: (a) foregrounding a broader conception of what constitutes work; (b) justice, in particular regarding historical and ongoing injustices between the Global North and South; and (c) autonomy and democracy as organising principles of a degrowth economy.

DEFINING DEGROWTH

Although it is most straightforwardly, albeit misleadingly, understood as material downscaling, degrowth denotes a far more encompassing transformation. Degrowth is indeed a proposal for voluntary, equitable and democratically led reduction of the materials and energy that a society extracts, processes and disposes of as waste.[2] Degrowth builds fundamentally on bioeconomics and ecological economics, which emphasise that biophysical limits to growth, in the form of resource availability or waste absorption capacity, are binding. In this sense, degrowth is a strong counter against visions of green growth and eco-modernisation, which rest fundamentally on claims of absolute decoupling, that is, delinking of economic growth from its biophysical impacts through the use and advancement of eco-efficient technologies. In various debates on the limits of eco-modernisation, degrowthers have demonstrated not only the lack of evidence for such delinking, but also cast doubt on its future likelihood, in particular its occurring at a pace and consistency that is required to avoid climate catastrophe.[3] This scholarship has also emphasised the rebound effects of eco-efficient technologies, the lower energy output-per-input of renewable energy sources and relatedly, the intense material requirements of eco-efficient technologies.[4] This call for downscaling, however, is not conceived as a technical matter of reduction, but rather an entry point for democratic process of societal decision-making on which activities to abolish, which ones to limit, and which activities to support and expand, that is, selective degrowth.[5]

Yet degrowth denotes a far more radical transformation that unsettles the dominant structures of our economies in more than one way. Firstly, it is more fundamentally a project to break with the dominance of economic growth as a societal goal, that is, the ideology of growth.[6] It is a call to deconstruct the automatic equation of more with 'better', or of economic growth with societal well-being in abstract terms, in order to open space for imagining other ideals and principles in organising economic relationships. Degrowth is rooted within a broader challenge to economism, that is, the power of economic rationality that dominates and smothers other social rationalities, goals and representations. This implies a radical questioning of economic imperatives such as efficiency and profit maximisation, and a (re)politicisation of the economy by challenging its supposed objective reality and foregrounding democratic choice in shaping it.[7] Degrowth is thus a project of reclaiming the economy.

Secondly, degrowth is not a quantitative issue of less (of the same), but rather a qualitative issue of 'different'. It is not contraction within a growth economy; rather, it denotes a reorientation of economic relations towards a different structure, in order to serve different functions.[8] It is a proposal to move towards a society in which the social metabolism – how societies organise their interaction with flows of materials and energy – is organised along different principles, such as needs and provisioning, care, solidarity, justice and democracy. This implies, most fundamentally, a structural-institutional change as economic institutions of capitalism, such as labour, welfare, property, markets, credit and public finance, perpetuate a growth imperative.[9] These institutions either depend on continuous economic growth for their functioning and sustainability (e.g. public services financing linked to growth through taxation systems, employment creation tied to economic expansion), or drive economic growth (e.g. interest-bearing credit, competition for greater market share). It also implies constructing and strengthening forms of production, exchange, labour, finance and consumption that are intentionally different from mainstream (capitalist) economic activity. Such alternative economic forms are more likely to prioritise production for concrete and situated needs, to foreground social and ecological values over accumulation, profit maximisation and growth, to localise production and consumption, and can cultivate values such as sharing, community, solidarity.

Cast this way, degrowth is first and foremost a project of restructuring and reorienting contemporary economic systems towards ones that centre needs and equitable provisioning rather than accumulation and economic growth. Such reorientation can take different paths: it implies a shift away from extractive activities, fossil fuel production, military and advertising, towards those that sustain and regenerate human and non-human well-being, such as healthcare, education, ecological-restorative agriculture and local food systems. It could mean, for instance, that subsidies and public financing provided to the former are eliminated and rerouted to the latter; that taxation systems are restructured in ways that punish harmful economic activity and reward life-sustaining activities; and that eco-social destruction created by capitalist growth economies is limited by establishing democratically determined caps on extraction of resources.

This restructuring/reorienting also implies ensuring equitable sustenance and access to basic goods and services for all, taking into account what access entails across different contexts and/or cultures. Among possible routes to this end are the decommodification of basic services such as healthcare, education and housing, and/or measures to guarantee a minimum level of well-being for all, for instance through universal basic income (UBI) or universal basic services (UBS) schemes. Guaranteeing a certain level of well-being for all through such arrangements is in fact a decision to radically shift how means of sustenance are distributed, that is, from being indexed to wage employment to being based on collectively assessed need. Delinking fulfilment of needs from employment status would not only relieve the coercion to work in exploitative, alienating and degrading jobs. It would also relieve the imperative to maintain economic growth for its employment creation potential.

GROUNDING DEGROWTH IN THREE AXES

There are three axes that are fundamental to degrowth's project of restructuring/reorienting that imbue it with its potential to unsettle dominant narratives of transition and join voices for radical eco-social transformation. I take them up in turn below.

Broadening 'work'

The first axis is a broader conception of what constitutes 'work' beyond commodity-producing wage labour, including the types of work that are fundamental for sustaining (human and non-human) life. Feminist thinkers have long theorised this domain of labour that falls outside of, yet underlies, commodity production, that is, social reproduction. Social reproduction is firstly the work of reproducing and sustaining labourers; but it also spans the production of life-sustaining goods and services and the regeneration of the social and ecological conditions of life and (commodity) production. Social reproduction thus includes not only the forms of labour that directly produce and sustain human capacity to produce, but also those that maintain, mediate and transform biophysical processes that undergird life.[10] What makes social reproduction particularly distinct is that it is markedly gendered (and racialised), on the one hand, and highly invisibilised and devalued, that is, codified as 'non-work', on the other. This is far from being accidental: commodity production under capitalism not only hides this sphere of work and production, but fundamentally depends on its devaluation: cheap, if not entirely free, production of labourers, their sustenance and the broader ecological-social conditions of production have been instrumental for the development and reproduction of capitalism.[11] Feminist scholarship has pointed to the global scale of the devalued and invisible value flows, drawing parallels between colonisation, domination of Nature and subjugation of women.[12] Social reproduction is thus global and includes the work of colonies, indigenous peoples and subsistence producers, which reproduce the global labour force and protect/regenerate natural metabolic cycles.[13] Added to this is the global division of social reproductive labour, where racialised social reproductive labour (e.g. of migrant care workers) serves to cheapen the costs of maintaining and reproducing capital accumulation, especially in countries of the Global North.

Foregrounding a broader conception of work entails, first of all, that this invisibilised sphere of labour and production is recognised, rewarded and supported. Possible actions to this end include implementing a care income, as well as expanding the rights and entitlements of essential workers and public investment into social and ecological reproduction. Such policies would not only provide material support for the workers of social reproduction but could also be instrumental in shifting perceptions of what is recognised and deemed valuable as work.

Yet recognition and validation are not sufficient for such a foregrounding. The mere recognition and validation of social reproduction, without problematising its organisation, risk perpetuating and solidifying its gendered (and racialised) distribution. A smaller social metabolism and downscaling of material and energy use carry with them important questions, such as what kind of activities will rely more on human labour, and whose labour will substitute for the reduction in energy use in, for instance, household production, agriculture or transportation. As feminist degrowthers have pointed out, given entrenched patterns of gendered division of labour, such structural shifts without ensuring gender justice runs the risk of re-feminisation of social reproduction.[14]

Crucially, feminist thinking and politics have not only been instrumental in pushing for recognising and rewarding the work of social reproduction. They have also problematised how this reproductive work is organised, that is, who will perform, how much of it, under which conditions and under whose control, if and how to remunerate it and how to decide on its distribution. In fact, for feminist politics, making social reproduction visible and to reveal it as work is not an end in itself, but rather the means for the struggle to alter its (gendered and racialised) distribution and the conditions under which it is performed. This is a critical insight, as it expands foregrounding a broader conception of work onto questions of how to organise social reproduction. Although there is hardly a blueprint, feminist scholarship and practice provide tools to tackle this question, pointing to cooperative and egalitarian forms of provisioning where labour is collective and organised along gender justice.[15]

To recap, degrowth's foregrounding of a broader conception of work is both a recognition and rewarding of the labour of social reproduction that is fundamental for sustaining (human and non-human) life, and a vision for its collective, egalitarian and democratic organisation. Such foregrounding provides a novel lens for thinking about transition justice, as it imbues not only the notion of transition but also that of justice with the diverse and immense field of labour and production that underpin commodity production and capital accumulation. That is to say, transition justice requires justice for (human and non-human) workers of social reproduction.

Degrowth as/through justice

The second fundamental axis is justice. Degrowth is a project of justice in two interrelated ways. Firstly, justice requires setting limits, as the social and ecological costs of growth are always unequally shared within and across societies and geographies. That is to say, downscaling of energy and resource use in itself is a project of justice. This is especially pertinent for the Global North–Global South relations, as economic growth in the North has been driving, and continues to drive, grave socio-ecological impacts on the South. It is therefore the North's responsibility to degrow, leaving more space for others to live.[16]

Secondly, and more importantly, growth is driven and enabled by global injustices. The unequal relationship between the Global North and the South, which is constituted historically and continues to be reproduced, lies at the basis of global capitalism. It positions countries of the North and

South differentially, where the former's prosperity and growth has been fundamentally dependent on the flows of cheap Nature and cheap labour appropriated from the latter. The historical dynamics of global capitalism that made the Global North wealthy have also put countries of the Global South on paths that have locked them into a perpetual growth imperative, for example, through structural dependency on extractivism, debt servicing or structural adjustment.

Repairing historical and ongoing injustices is thus fundamental to degrowth, and equips it with a crucial international dimension. While degrowth is predominantly a proposal developed in and for the core-industrial countries of the Global North, with its associated policies and actions often envisioned as interventions within these economies, the implications of the 'responsibility to degrow' are by no means limited to the geographical boundaries of the Global North. That is to say, degrowth as justice is necessarily a project of addressing historical and contemporary impacts of economic growth, on the one hand, and the growth-reproducing structures of the global economic system, on the other.

Such recasting of the link between degrowth and justice is indeed central in recent degrowth thinking and activism, crystallised especially around the notions of ecological debt, that is, the historical and contemporary appropriation and/or disproportionate use of ecological resources and sinks, and ecologically unequal exchange, that is, unequal flows of embodied Nature through goods traded in international trade.[17] Yet this needs to be complemented with the global perspective on social reproduction, which expands this notion of justice to include unequal flows of life-sustaining labour of humans and Nature between the Global North and the Global South. Seen this way, it is not only the flows of (embodied) Nature, either through direct use and appropriation or unequal exchange in global trade, but more broadly flows of social reproductive labour that sustains and reproduces capitalist growth. Actions towards repairing global injustices should therefore take into account a broader notion of 'social reproductive debt' that includes the racialised and cheapened social reproductive labour flowing from the Global South to the North, as well as colonial reparations and giving land back to their rightful indigenous custodians.

The concrete actions and interventions that emerge from this particular understanding of degrowth as/through justice can be broadly categorised under three headings, which are widely congruent with the proposals made in the chapter on debt in this book. The first pertains to repairing historical and contemporary injustices and includes measures such as repayment of ecological and, more broadly, social reproductive debt, climate and colonial reparations, and interventions in the global financial and trade system that reverses/alleviates dynamics of unequal exchange between countries of the Global North and the South. In this sense, degrowth does not only join contemporary movements that call for reparations and indigenous sovereignty like the Land Back Movement,[18] but also those that revive the transformational potential of the Southern Peoples' Ecological Debt Creditors Alliance which had reframed the so-called debt crisis of the Third World in terms of the debt owed by the Global North.[19]

The second set of actions/interventions relates to the potentially debilitating impacts that the contraction of production and consumption activities in line with degrowth in industrialised countries would have on the Global South, especially on countries that are structurally dependent on export or foreign investment.[20] As the asymmetric relationship and the unequal flows and Nature and labour between the Global North and the South has also historically shaped many economies in the South to be structurally dependent on export sectors, the latter would suffer in the case of a contraction in the North, amounting to a coerced delinking. Although justice-oriented measures mentioned above would provide some relief, direct measures such as transfer of resources for economic restructuring are also called for.

And the third and final set of proposals is about opening and strengthening the space for the Global South to pursue non-growth pathways if it chooses to do so. This implies recognising the validity of the variety of movements, proposals and worldviews beyond growth originating from the Global South (e.g. post-extractivism, Ubuntu, Buen Vivir) on the one hand, and measures to relieve the built-in imperative of growth in the Global South by, for instance, financing cooperative/public systems of provisioning delinked from growth or supporting a shift away from dependency on unequal exchange relations, on the other.

Degrowth as autonomy/democracy

The third and final axis is autonomy and democracy. This relates to degrowth's call to exit a social imaginary dominated by the imperative of growth, and to foreground democratic decision-making in shaping economic processes. A counterpart to this call has been degrowth's emphasis on autonomy. Degrowth is inspired heavily by the conceptions of autonomy (and, relatedly, democracy) developed by thinkers such as Ivan Illich, André Gorz and Cornelius Castoriadis. Despite their differences, the common ground shared by these thinkers is an understanding of how the increased scale of economic activity undermines the ability to self-govern, be it through the centralisation and bureaucratisation of economic decision-making or the erosion of the ability to self-define needs with the rise of the market economy. That is to say, endless economic growth is not desirable, even if it was biophysically possible, as it displaces the ability to collectively self-govern.

Democratising economic decision-making towards expansion of self-governance, that is, enabling all to participate in the making of decisions that affect their lives, is therefore inherent to degrowth. This is animated, firstly, in degrowth's insistence of collective and democratic determination of situated needs and limits, that is, which activities to abolish, which ones to limit, and which activities to support and expand in a degrowth future. But it also resonates with degrowth's emphasis on '*different*, not only less', that is, its call for constructing a different kind of economy that serves functions that are different than one that is built on exploitation, accumulation and growth, towards one that centres needs, provisioning, equity and solidarity. Curbing corporate power, establishing democratic oversight over money and finance, participatory public budgeting, democratic governance of productive capacities as well as constructing and strengthening alternative (non-capitalist) forms of production, distribution/exchange and consumption are thus fundamental facets of degrowth.

Democratisation of economic decision-making at various scales through such interventions and practices has the potential to foreground concrete needs, use values and non-monetary wealth over accumulation, profit maximisation and growth, and prioritise principles such as ensuring sustainable and equitable livelihoods or regeneration, renewal and protection of environmental quality.[21] Opening economic decision-making processes to democratic participation of a wider base of actors would enable the involvement of a broader range of demands and values in informing decisions regarding, for instance, what, how much and for whom to produce under which conditions, how to set prices or wages and where to invest surplus. This would open space to rethink economic imperatives such as growth or efficiency, enable the operationalisation of alternative goals, and would (re) politicise the economy by subjecting economic rationality to societal deliberation and control.

That is to say, democracy and autonomy within the economic realm are not only principles worth pursuing in themselves, but they would also function as a force to curb and transform the socially and ecologically destructive dynamics of capitalist growth economies. Degrowth's emphasis on economic democracy and autonomy is particularly critical against the backdrop of mainstream debates on eco-social transition. The proposals within that front mostly centre on a structural reorientation of economic activities, such as shifting away from fossil fuel-based sectors, often coupled with the use of eco-efficient technologies. They reduce the question of transformation to one of getting the investments 'right', that is, away from ecologically destructive activities and correcting the misallocation of productive capacities. Missing from these debates, however, is a vision of how economic processes are to be governed and what kind of economic institutions are needed. This is where degrowth's emphasis on autonomy/democracy becomes crucial, as it equips debates on transition with a problematisation of the processes of economic decision-making, in addition to their outcomes.

CONCLUSION

Degrowth is predominantly a proposal developed in and for the core-industrial countries of the Global North. It is not a blueprint nor a vision to be imposed on the rest of the world, but rather one among many other visions of living well and equitably beyond capitalist growth, which often have preceded degrowth. Degrowth is also not homogenous, static, or without contradictions, much like other social movements. A predominance of self-limitation, right sizing and an exclusive focus on consumerism, for instance, cannot be denied especially in earlier strands of degrowth thinking and practice. A tendency to shy away from engaging with the global justice implications and responsibilities of degrowth can also be discerned.

Yet a more direct engagement with capitalism, in particular its global ecological regime, colonialism and patriarchy, on the one hand, and anti-capitalist forms of organising the economy, on the other, have been taking place in degrowth thinking, not least due to its broader taking up by social movements and its opening up to be located next to multiple visions that challenge the hegemony of growth and capitalism. This turn has imbued degrowth with the very foundations that are essential in its unsettling and challenging the dominant understandings of transition, and points to the need for degrowth to keep evolving as a living political project by the many worlds that fit a world.

NOTES

1. M. Bosquet (André Gorz). *Nouvel Observateur*, Paris, 397, 19 June 1972, p. IV. Proceedings from a public debate organised in Paris by the Club du Nouvel Observateur.
2. Federico Demaria, François Schneider, Filka Sekulova and Joan Martinez-Alier. 'What is degrowth? From an activist slogan to a social movement'. *Environmental Values* 22(2) (2013): 209.
3. Jason Hickel and Giorgos Kallis. 'Is green growth possible?' *New Political Economy* 25(4) (2020): 469–86.
4. Giorgos Kallis. 'Radical dematerialization and degrowth'. *Philosophical Transactions of the Royal Society A: Mathematical, Physical and Engineering Sciences* 37(2095) (2017): 20160383.
5. Giorgos Kallis. 'In defence of degrowth'. *Ecological Economics* 70(5) (2011): 875; Ulrich Brand, Barbara Muraca, Éric Pineault, Marlyne Sahakian, Anke Schaffartzik, Andreas Novy, Christoph Streissler et al. 'From planetary to societal boundaries: an argument for collectively defined self-limitation'. *Sustainability: Science, Practice and Policy* 17(1) (2021): 264–91.
6. Serge Latouche. *L'Invention de l'Économie* (Paris: Albin Michel, 2005).
7. Valérie Fournier. 'Escaping from the economy: the politics of degrowth.' *International Journal of Sociology and Social Policy* 28(11/12) (2008): 535–37.
8. Giorgos Kallis, Federico Demaria and Giacomo D'Alisa. 'Introduction: degrowth', in Giacomo D'Alisa, Federico Demaria and Giorgos Kallis (eds), *Degrowth: A Vocabulary for a New Era* (London: Routledge, 2014), p. 4.
9. Kallis, 'In defence', 877.
10. Stefania Barca. 'The labor(s) of degrowth'. *Capitalism Nature Socialism* 30(2) (2019): 214.
11. Maria Mies. *Patriarchy and Accumulation on a World Scale: Women in the International Division of Labour* (London: Zed Books, 1986).
12. Mies, 'Patriarchy'; Silvia Federici. *Caliban and the Witch: Women, the Body and Primitive Accumulation* (Brooklyn, NY: Autonomedia, 2004).
13. Maria Mies and Vandana Shiva. *Ecofeminism* (London: Zed Books, 1993); Ariel Salleh, *Ecofeminism as Politics: Nature, Marx and the Postmodern* (London: Zed Books, 2017).
14. Anna Saave and Barbara Muraca. 'Rethinking labour/work in a degrowth society', in Nora Räthzel, Dimitris Stevis and David Uzzell (eds), *The Palgrave Handbook of Environmental Labour Studies* (Cham: Palgrave Macmillan), pp. 743–67.
15. Silvia Federici. *Re-enchanting the World: Feminism and the Politics of the Commons* (Oakland, CA: PM Press, 2019).
16. Susan Paulson, Giacomo D'Alisa, Federico Demaria and Giorgos Kallis. *The Case for Degrowth* (Cambridge: Polity Press, 2020).
17. Matthias Schmelzer, Andrea Vetter and Aaron Vansintjan. *The Future is Degrowth: A Guide to a World beyond Capitalism* (Brooklyn, NY: Verso Books, 2022).

18. 'Land Back: A Yellowhead Institute Red Paper', https://redpaper.yellowhead institute.org (last accessed May 2023).
19. Elizabeth Bravo and Ivonne Yánez (eds). *No More Looting or Destruction! We the Peoples of the South are the Ecological Creditors* (Quito: SPEDCA, 2003), https://digitalrepository.unm.edu/cgi/viewcontent.cgi?article=1398&context= abya_yala.
20. Corinna Dengler and Lisa Marie Seebacher. 'What about the Global South? Towards a feminist decolonial degrowth approach'. *Ecological Economics* 157 (2019): 248–49.
21. Nadia Johanisova and Eva Fraňkovà. 'Ecosocial enterprises', in Clive Spash (ed.), *Routledge Handbook of Ecological Economics: Nature and Society* (New York: Routledge, 2018), pp. 513–15.

15

Degrowth, Climate Emergency and the Transformation of Work

Luis Gonzalez Reyes

TWO CHALLENGES AND A CONDITIONING FACTOR FOR ECO-SOCIAL TRANSITIONS

Any eco-social transition faces two fundamental challenges. The first is the degree of depth and breadth of the changes required. The second is the speed at which those transformations need to happen.

As to the depth and breadth of the necessary changes, we can highlight three fundamental aspects that require change. The first is our energy matrix, which must shift from being based on fossil fuels to renewable energies. This is the only way to avoid the worst-case scenarios of possible climate emergency or mass extinction of species. This is not a small change, since renewable energies have different characteristics than fossil-based energy. Specifically, they are low-concentrated sources that function as irregular flows and generate a significantly lower amount of available energy.[1] To think of a world driven by renewable energies is to think of a different economy and society.[2]

The second factor is that not only fossil energy sources are running out;[3] the same thing is happening to certain elements. This is true for elements like phosphorus, which is essential in industrial agriculture, as well as many elements that are central to high-performance renewables.[4] In other words, the energy transition also has to move toward simple renewable technologies and materials. Changing the material bases of our economy implies a transition from economies of extraction (mining) to economies of production, which are none other than agriculture. This includes moving from economies based on minerals to others that are centred around biomass.

A change in the economic model is also essential, since capitalism requires constant growth to avoid falling into crisis (that is, for it to be able to function); such constant growth is impossible to maintain. Among other factors, this is due to the fact that, at the global level, there is a linear correlation between GDP and material and energy consumption.[5] In other words, empirical data reflects that there is no increase in GDP without an increase in material and energy consumption.[6] There is no dematerialisation of the economy.

We now will address the second challenge: the speed of the transition. One of the elements forcing an accelerated transition is the climate emergency. Climate change is not a linear process. After reaching a certain threshold (which is probably a temperature increase of 1.5°C, which is already close at hand), the planet itself will become a net emitter of greenhouse gases. None of this will stop until a new equilibrium is reached, which will be 4–6°C higher than pre-industrial temperatures. Such a result would make the vast majority of the Earth uninhabitable for humans.[7] Something similar happens to ecosystem dysfunction.[8] Therefore, in response to the climate emergency, we need to determine what the reduction in emissions should be so as not to exceed 1.5°C. The United Nations proposes that globally, this reduction must be 7.6 per cent per year.[9] This implies a 58 per cent reduction in 2030 compared to emissions in 2019. However, in a world marked by inequality, the responsibilities between some territories and others are very different. For the world's top historical and per capita emitters, reductions would need to be greater, in the range of at least 10 per cent per year. This means a 65 per cent drop in 2030. To give an idea of what this implies, the reductions that occurred in the former USSR when it collapsed were in the order of 4 per cent during the years with the largest drops. In this case, we are talking about an annual rate that is almost double that, but at a planetary level and sustained over time.

This has a corollary, which is that we cannot carry out programs in two stages and apply a strategy that can be summed up as 'first do the easiest to save time and then move on to the difficult parts'. The change has to happen in a single step. It needs to be done all at once.

Addressing these challenges entails a radical transformation in the world of work and, more specifically, employment. In this chapter, I suggest that we need to perform four transformations when it comes to jobs:[10]

i.   First, encourage restoration, knowing that these are occupations that, if done well, will gradually disappear.

ii. Second, encourage occupations that take care of and integrate into the environment and, by doing so, generate more jobs of this type. This produces a positive feedback loop. Agroecological agriculture is an example. This is the determining area for action.

iii. Third, reconvert activities that produce services for the economy that are harmful to life and whose demand increases as biodiversity is depleted. One example is the production of chemical fertilisers. These are very dangerous, because they create the illusion that we are not eco-dependent.

iv. Fourth, reconvert activities that depend on good ecological status but are based on environmental exploitation. One example is intensive fishing, but also the banking services that support it. These occupations are self-regulating, as they automatically disappear without a healthy environment. However, it is essential to act before that happens.

Thus, the changes required are profound and must occur very quickly. At the same time, such changes are inevitable. Let's not delude ourselves into thinking that we can avoid making them: a degrowth, localisation and primarisation of the economy will occur as a consequence of environmental limits. For example, without abundant oil, it is not possible to maintain the globalised production and consumption system, or a highly urbanised population.[11]

What is at stake is how fair the transition is, and how much degradation occurs before it happens, not the ecological transition itself. With all this, what are the major lines of transformation needed?

REDUCTION OF MATERIAL AND ENERGY CONSUMPTION TO BE WITHIN ECOLOGICALLY VIABLE LIMITS

If we want to avoid the worst consequences of the collision with ecological limits, it is essential to develop a robust contraction of the sphere of production: a degrowth of the social metabolism (see Bengi Akbulut's chapter). This contraction must happen now. In the 1970s or 1980s, a period of growth fuelled by high-tech renewable energy might have been feasible, followed by an inevitable decline in metabolism to bring it within the framework of what is physically and ecologically sustainable. Today, in the 2020s, there is no longer time for that if we are to have any chance of keeping climate change from skyrocketing, to avoid exceeding a 1.5°C increase in temperature.[12] Furthermore, as we noted, there are probably no energy or material resources for this industrial development.

In Spain, we have modelled the changes in the economy that would be needed to bring us within the ecological safety margins in the 2020–2030 decade. To do this, we took the Spanish economy as a whole (productive and reproductive) broken down by activities, and then translated these activities into hours of work and emissions. Based on the 2019 data, we applied a series of policies that caused some activities to grow and others to shrink in activity and emissions between 2020 and 2030. Finally, we translated the resulting work hours in 2030 into jobs (in the case of salaried hours) and compared the emissions to what the UN is proposing as necessary to avoid a 1.5°C temperature increase (average global reduction of 7.6 per cent in CO_2 emissions),[13] while also contemplating the Spanish ecological debt (annual reduction of 10 per cent for Spain).[14]

In policies, we define a *Green New Deal scenario* based on the massive deployment of high-tech renewables and ICT, but also on the contraction of mobility, a decrease in heating and air conditioning and the development of organic farming. This scenario would reduce emissions considerably in the decade under study (-45 per cent), but far from sufficiently (-65 per cent), even without considering the ecological debt (-58 per cent). Things are not moving along quickly enough, and speed is essential in the climate emergency scenario, since the longer it takes to get the concentration of CO_2 below 350ppm (currently it is well above 410ppm), the more likely it is that the 1.5°C threshold will be exceeded.

We have also modelled out a *degrowth scenario*, which differs from the Green New Deal scenario, and would achieve the robust emission reduction needed within the decade (-68 per cent). To visualise the level of economic activity, in 2030 this would be somewhat lower than what existed in April 2020 in Spain during the strictest part of the COVID-19 lockdown.

This degrowth economy would imply an increase in activity in some sectors, such as waste management or food, which would create jobs.[15] Much emphasis has been placed on these green employment options from environmental and union sectors, sending the message that ecological transitions would generate an opportunity in the job market. However, we need to look at the whole picture, including those sectors that have an excessively large size and/or cause ecological destruction. Upon analysing the labour market as a whole, a green transition shows a significant net reduction in hours of productive work, at least in the short term.[16] This is a challenge that makes the eco-social transition much more complicated and requires more than just environmental measures, as we will discuss further.

Which sectors would require a decrease and which would have to increase in a degrowth transition? If we analyse the different productive sectors, the hours of work dedicated to construction, transportation, finance, tourism, industry and ICT would require a significant decrease. In contrast, the energy sector and, above all, forestry and food, would experience major increases. All of this would also be subject to a major reconfiguration, as we will explore further.[17] For the Spanish case, these results will likely be qualitatively equivalent to the rest of the economies of the core countries.

The emissions arising from heating and air conditioning in public and private spaces would have to be significantly reduced. Beyond measures to increase efficiency, this would imply changing air conditioners to fans, or going from heating full homes to heating only certain rooms (the bathroom or the living room) or people (brazier heaters under tablecloths). It would also require a major reduction in transportation by plane and car.

RELOCALISATION AND DIVERSIFICATION OF THE ECONOMY

The current globalised model accumulates ecological and social impacts that are incompatible with sustainability. In addition, to put an end to the inequality and extractivism that make today's globalisation possible, it is necessary to relocalise economies to make sustaining life in one territory function based on nearby resources. However, the imperative goes beyond this, as there is no substitute for oil that makes it possible to maintain the current model of transporting large volumes and masses, over long distances and at high speeds, of people, merchandise and information.[18] In reality, relocalisation is something that will happen as oil becomes increasingly scarce. This relocalisation necessarily implies a diversification of economic activities so that territories are capable of meeting most social needs on their own.

An example of this relocalisation of the economy is the model we developed for Spain, in which we propose an 80 per cent decrease in maritime traffic in 2030 as compared to 2020 (which is the way most goods enter the majority of central economies).[19]

Diversification would be most important in the industrial sector, which is also one of the sectors that would most have to undergo profound changes. The transformations in this sector must encompass a three-fold change. First, a much greater diversity of the productive fabric in order to cope with a less globalised economy that continues to meet people's needs. This will take different forms depending on the productive specialisations of each territory, but it is likely that a revitalisation of food processing, furniture manufacturing or textiles will be quite common. Second is a reduction in industries with major environmental impact: pesticides, automobiles, and, unfortunately, a long list of etcetera. Third is a transformation towards low environmental impact. This implies a profound transformation of the entire energy sector, starting with the technologies it uses, which would not be based on fossil fuels or minerals, but rather on truly renewable energies, biomass and abundant, easily extractable and reusable materials.

INTEGRATION OF SOCIAL METABOLISM WITHIN THE ECOSYSTEMIC METABOLISM

Ecosystems are much more powerful and resilient than industrial capitalism. This means that if ecosystems focus their efforts not on growth but on closing cycles (for example, recycling carbon, nitrogen or phosphorus at rates of 99.5–99.8 per cent), using solar energy, maximising diversity and with high degrees of cooperation,[20] human economies should try to do the same. This means that societies should dedicate the bulk of their productive effort to the primary sector under the agroecological paradigm, since economies focused on the secondary or tertiary sectors are not capable of satisfying the essential closed cycles in ecosystems.[21]

One way of looking at this more concretely is the energy model. Energy would evolve from being a mix based mostly on imported fossil fuels to a renewable option. However, not as it is usually conceived: with high-tech renewables built with non-renewable materials and energy and used primarily to produce electricity (which accounts for approximately 20 per cent of global energy consumption). The transition would rather be towards truly renewable energies and of an emancipatory nature (see the chapter by Pablo Bertinat and Tatiana Roa Avendaño in this book). These have different characteristics: i) they are built with renewable materials and energy; ii) they integrate and take advantage of the functioning of ecosystems (such as the bioclimatic heating of a house or the use of stable air currents in the oceans for transport); iii) they do not monopolise all the energy flow for humans, but leave it for the rest of living beings, which entails less energy accumulation and intermittent use; iv) they do direct work (grind, beat, pump, etc.) and produce heat, not just electricity; v) they are controlled by the community.[22]

However, the integration of social and ecosystem metabolism is likely most appreciated in the food sector, where agriculture needs to be fostered using green manures, pest control based on ecosystem balances, short marketing circuits, adaptation of crops to climatic and soil conditions, productive diversity, seed control, food sovereignty, etc.

INTEGRATION OF PRODUCTION AND REPRODUCTION INTO A SINGLE ECONOMIC UNIT

To start with, we should not separate reproductive work from productive work, since both are inextricably linked, because reproductive work undergirds the possibility for all production work to exist (for a deep dive into this debate, see the chapter by Akbulut in this book). Furthermore, one of the roots of patriarchy is the division of reproduction and production, and specialisation by gender, something that is still maintained both in central and peripheral regions: women dedicate more hours to reproductive tasks than men, which in total results in more working hours.[23] Thus, a first major idea is the necessary integration of production and reproduction in the same economic unit, as in the case of a peasant family, but with an equitable distribution of all tasks between genders. In this example, a typical peasant family would no longer serve as a reference.

A second factor that results from contemplating work, both reproductive and productive, is that we dedicate more hours of the day to reproductive work in Spain,[24] and it is very likely that this situation will be repeated in the rest of the world.[25] Thus, thinking about transitions in the world of work forces us to focus our attention on what happens at home and is essential for social reproduction, despite this normally being kept in a black box.

Add to these two elements the fact that we must revolutionise the current social value placed on jobs. Today, it is the productive jobs, and specifically those situated in the control centres, that facilitate the reproduction of capital, which are the ones that receive the highest social (and economic) value. From a point of view that assumes our interdependence and eco-dependence, care work should be considered essential and, therefore, more valuable.[26] The proposal is that these care jobs grow not only in value (social, not monetary),[27] but also in dedication, assuming as part of the decommodified work on a communal basis, as we will delve into a little later.

FORCE A STRONG REDISTRIBUTION OF WEALTH BETWEEN AND WITHIN TERRITORIES WITH CRITERIA OF GLOBAL JUSTICE

As noted above, an economy that fits within the limits of ecosystems implies a smaller economy with fewer jobs. This is a tragedy in societies marked by cross-cutting, deep-seated inequalities, in which there are significant pockets of impoverished people and in which the dependence on wages to meet needs is very high. For this reason, the ecological transition must be accompanied by a far-reaching social transition articulated around the redistribution of wealth.

This would necessitate measures such as expropriations of large estates, introduction of basic income, strong redistributive tax policies, etc. At the international level, we would speak of restitution of ecological and colonial debts, as discussed by Alberto Acosta, Miriam Lang and Esperanza Martínez in their chapter in this book. Within the labour framework, the distribution of employment is central. In the models that we have developed for Spain, both to address climate change[28] and the loss of biodiversity,[29] under the current labour framework, we have a net destruction of employment, but with 30- or 32-hour work weeks, there is net creation. These results could be extrapolated qualitatively at a global level. Thus, the reduction of the working day, the prohibition of overtime, or an earlier retirement age appear to be the policies central to a just transition.

Organised society can be an active agent of this redistribution without intervention by the state, for example, squatting in homes and on land, or forcing reductions in working hours without losing any salary. In any case, the state should also be forced to be part of the process. But the idea is that these struggles, which are very costly, should not last indefinitely; rather, the population should be able to sustain their lives autonomously, without depending on the market or the state. This is what the last idea addresses.

INCREASED ECONOMIC AUTONOMY OF ALL PEOPLE

One of the pillars on which capitalist societies are based is the lack of economic and material autonomy of people, which is actively destroyed through processes of accumulation by dispossession.[30] Once land has been expropriated, the territory has been degraded and communities dissolved, the collective self-management of basic aspects of survival such as food, shelter or clothing become impossible. Lacking this autonomy, the population has no choice but to get much of what they need on the market, for which they require money. And for most, that means working for a salary. From then on, we voluntarily or involuntarily become accomplices in sustaining the system on which our subsistence depends.

Additionally, the proper functioning of this system requires permanent growth that becomes an irrational obligation. Growth occurs within the system and at the expense of those existing outside of it. Internally, capitalism produces to generate surplus value that has to be reinvested to continue expanding capital, not to offer goods and services that meet social needs. This compulsive growth is one of those responsible for ecological devastation, as shown by the aforementioned linear relationship between energy consumption, material consumption and GDP worldwide. This correlation also occurs between the concentration of CO_2 in the atmosphere and GDP.[31]

Added to the above is the fact that capitalism needs to constantly expand to areas where it does not rule the day. On the one hand, this capitalises on human work that existed outside the markets, invading more and more territories (accumulation by dispossession, as we mentioned before). A historical example of how the system converts external jobs into capital has been the enslavement of the African population to generate surplus value in the American sugarcane or cotton plantations. Another area into which it expands is the conversion of non-human 'work' into capital.[32] One example is oil. By extracting and turning it into a commodity, the 'work' of millions of years of concentration, compression and heating of huge amounts of organic matter is capitalised until turning it into a high-density energy source. One last area of expansion of the system with which capital manages to reproduce itself is by introducing new facets of our lives that were not previously commodified into this mercantile logic; examples would be care for the elderly or our social relations, today in the hands of digital companies. All this generates eco-social destruction.

Thus, the proposal consists of moving from 'market' societies (in which the satisfaction of needs is mostly achieved through the purchase of goods and services), towards societies 'with markets, some of them regulated' (in which the population has a high level of economic autonomy). In these economic orders, the basic goods and services for subsistence that were commodified would have their access price regulated to guarantee that it is universally accessible.

This implies fundamentally profound processes of *desalaryisation* and *decommodification*, and a decisive move towards cooperative and self-managed forms of ownership and production. The objective of a degrowth economy is not simply to 'dignify' the conditions of wage labour or to sustain and expand welfare states. To achieve said transformations, it will not be enough to reassign 'jobs', even if they are 'green': it will also be necessary to break down the mechanism of wage labour as the fundamental pillar that organises social relations.

To do this, advancing in the social control of the means of production is key in a process of *desalaryisation* and removing from the market more and more activities that *decommodify* our lives. This means defending and rebuilding common goods that allow for a new organisation of subsistence that not only removes itself from mercantile dynamics, but also reappropriates all the autonomous decision-making capacity that has been expropriated by the state. We need to make territory, but also rights, care and education into common goods that are once again in the hands of the people and thus, can detach themselves from the destructive dynamics of capitalism and submit to democratic decision-making.

FINAL REMARKS

Carrying out this type of transition transcends the trade union sphere, even if trade unionism were to seriously focus its efforts on the task of building worker autonomy based on cooperativism. This means combining visions and struggles with environmentalism, feminism, internationalism or cooperativism, requiring a holistic perspective and action.

It also implies the necessity of giving new meaning to the social concept of work in various ways. First, separating it from employment, and extending it to care work and productive community work. Second, socially devaluing employment and fighting to destroy it. Third, betting only on jobs at the service of social reproduction and the fabric of life, and not having a focus on the reproduction of capital. If used at all, they should enable the satisfaction of human needs. And fourth, breaking with the productivist view of work and redefining it as a source of personal and collective meaning that does not come into conflict with leisure. For example, if the measures proposed in our model are implemented,[33] we would work fewer hours in total, dedicate more time to unpaid care and less to employment (both public and private), in a self-managed non-capitalist work field that is framed in the feminist, ecological and solidarity economy. This would likely lead us to a life that is closer to living well or living a flavourful life.

NOTES

1. Luis González Reyes. 'Crisis energética'. *Papeles* 156 (2022).
2. Andreas Malm. *Capital fósil* (Madrid: Capitan Swing, 2021); Ramon Fernández Durán and Luis González Reyes. *En la espiral de la energía* (Madrid: Libros en Acción, Baladre, 2018).

3. Antonio Turiel. Petrocalipsis. *Crisis energética global y cómo (no) la vamos a solucionar* (Madrid: Alfabeto, 2021).
4. Alicia Valero and Antonio Valero. *Thanatia. The Mineral Limits of the Planet* (Barcelona: Icaria, 2021).
5. Gail Tverberg. 'The world's energy problem is far worse than we're being told', oilprice.com; José Bellver. 'Costes y restricciones ecológicas al capitalismo digital'. *Papeles* 144 (2019).
6. Timothée Parrique, Jonathan Barth, François Briens, Christian Kerschner, Alejo Kraus-Polk, Anna Kuokkanen and Joachim H. Spangenberg. *Decoupling Debunked: Evidence and Arguments against Green Growth as a Sole Strategy for Sustainability* (Brussels: European Environmental Bureau, 2019).
7. David I.A. McKay, Arie Staal, Jesse F. Abrams, Ricarda Winkelmann, Boris Sakschewski, Sina Loriani, Ingo Fetzer, Sarah E. Cornell, Johan Rockströmand and Timothy M. Lenton. 'Exceeding 1.5°C global warming could trigger multiple climate tipping points'. *Science*, 9 September 2022, DOI: 10.1126/science.abn7950.
8. IPBES. 'El informe de la evaluación mundial sobre la diversidad biológica y los servicios de los ecosistemas'. IPBES, 2019.
9. UNEP. 'Emissions Gap Report 2019', 2019.
10. Jean-François Ruault, Alice Dupré la Tour, André Evette, Sandrine Allain and Jean-Marc Callois. 'A biodiversity-employment framework to protect biodiversity'. *Ecological Economics* 191 (2022), DOI: 10.1016/j.ecolecon.2021.107238.
11. Durán and Reyes, 'En la espiral'.
12. Jaime Nieto, Óscar Carpintero, Luis J. Miguel and Ignacio de Blas. 'Macroeconomic modelling under energy constraints: global low carbon transition scenarios'. *Energy Policy* 137 (2020), DOI: 10.1016/j.enpol.2019.111090.
13. UNEP. 'Emissions Gap Report'.
14. Luis González Reyes et al. 'Escenarios de trabajo en la transición ecosocial 2020–2030'. Ecologistas en Acción, 2019.
15. Ángeles Cámara and Rosa Santero-Sánchez. 'Economic, social, and environmental impact of a sustainable fisheries model in Spain'. *Sustainability* 11–22 (2020): 6311, DOI: 10.3390/su11226311.
16. See Antal, 2014; Bowen and Kuralbayeva, 2015; González Reyes et al., 2019; Nieto et al., 2020; Otero et al., 2023.
17. Gonzáles Reyes et al. 'La transicion ecosocial'; E. Oteros, C. Monasterio, A. Gutiérrez, M. Hernández, I. Álvarez, D. Albarracín, L. González Reyes, J.L. Fernández Casadevante, G. Amo, M. García, V. Hevia, I. Iniesta and C. Quintas (2023). 'Biodiversidad, economía y empleo en España. Análisis y perspectivas de futuro'. Amigos de la Tierra, Ecologistas en Acción, SEO BirdLife, WWF, Madrid.
18. Ignacio de Blas, Margarita Mediavilla, Iñigo Capellán-Pérez and Carmen Duce. 'The limits of transport decarbonization under the current growth paradigm'. *Energy Strategy Reviews* 2020, DOI: 10.1016/j.esr.2020.100543.
19. Gonzáles Reyes et al. 'La transicion ecosocial'.
20. Carlos de Castro. *Reencontrando a Gaia* (Malaga: Ediciones del Genal, 2019).
21. Circle economy. 'The circularity gap report 2022', Circle economy, 2022.

22. Reyes, 'Crisis energética'.
23. ONU. *Informe de los Objetivos de Desarollo Sostenible 2016* (New York: UN, 2022).
24. Gonzáles Reyes et al. 'La transicion ecosocial'.
25. ONU. *Informe.*
26. Amaia Pérez Orozco. *Subversión feminista de la economía. Aportes para un debate sobre el conflicto capital-vida* (Madrid: Traficantes de Sueños, 2014).
27. Corinna Dengler and Miriam Lang. 'Commoning care: feminist degrowth visions for a socio-ecological transformation.' *Feminist Economics* (2021), DOI: 10.1080/13545701.2021.1942511.
28. Gonzáles Reyes et al. 'La transicion ecosocial'.
29. Oteros et al. 'Biodiversidad'.
30. David Harvey. *El nuevo imperialismo* (Madrid: Akal, 2007).
31. Tim Garret. 'What is your carbon footprint?, 2018, www.inscc.utah. edu/~tgarrett/what-is-your-carbon-footprint.html inscc.utah.edu.
32. Jason W. Moore. *El capitalismo en la trama de la vida. Ecología y acumulación de capital* (Madrid: Traficantes de Sueños, 2020).
33. Gonzáles Reyes et al. 'La transicion ecosocial'.

16

Nayakrishi Andolon: Alternatives to the Modern, Corporate Agri-Food System in Bangladesh

Farida Akhter

INTRODUCTION

The people of Bangladesh are 'victims' of the predatory foundation of colonial-industrial civilisation based on fossil fuel. The country as a site of industrial extractive expansionism is exceptionally vulnerable to climate change. It has been facing extreme weather with more and more frequent natural disasters like storms, cyclones, ocean surges, drought, erosion, landslides, flooding and salinisation. Estimates show that by 2050, one in every seven people in Bangladesh will be displaced by climate change and up to 18 million people may have to move because of sea-level rise alone.[1] Added to this is the displacement due to development policies imposed by multilateral and bilateral institutions. Violent population displacement from the agrarian economy and rural livelihood is systematic. The inherent process of capitalist transformation alienates people from land and ceaselessly forces them to migrate to cities to become cheap labour in the ready-made garment sector.

As a country frequently hit by disaster, the people of Bangladesh have developed a rich practice of disaster management. People's active participation in recovering from disaster generates a form of collectivism that enhances the resilience and the survival potential of the affected community. People build networks, activate old relations, and come forward for mutual aid, demonstrating the power of self-determination and localisation of people's power. However, this potential of the people has never been harnessed. Instead, a highly coercive state imbued with centralised power has copied and imposed laws, structures and administrative culture from the colonial era. International development actors that have a narrow and technical understanding of disaster recovery have imposed highly bureaucratic and predesigned disaster aid and development policies. These constitute massive challenges for the people.

The just transition in Bangladesh, therefore, implies two interrelated but parallel strategies:

1. Systematic critique and resisting the idea of 'development' rooted in colonial structures and capitalist model of industrial civilisation. The primary requirement is the appropriate redesigning of economic, social, cultural and technological transformation by ecological principles.
2. To design appropriate strategies for just eco-social transition in the industrial and agrarian sectors.

A systematic understanding of options available for food production, labour utilisation and the development of knowledge, skill and productivity must have local features rather than models imposed from outside.

A just, equitable and sustainable transition cannot happen by letting the unjust system run as usual. The massively destructive extractive system became more visible after climate change and the COVID-19 pandemic. To get out of the extractive system, just transition has to ensure justice for both human beings and the environment. Nayakrishi Andolon, the farmers' biodiversity-based ecological movement, is a new pathway to achieve this goal.

Based on decades of collective reflection and praxis with Bangladeshi NGO, UBINIG, this chapter offers a critique of modern and corporate agrifood system in Bangladesh as well as foregrounds an alternative solution and movement from below, the Nayakrishi Andolon. It traces the origins of Nayakrishi and foregrounds the praxis of biodiversity-based farming as viable and better agricultural system for the people and the planet.

Through its praxis of women-led seed networks and knowledge practices, Nayakrishi demonstrates the possibilities and power of indigenous peoples, women and farmers as knowledge bearers who are at the frontlines of Bangladesh's agri-food system. By offering this concrete case, the chapter stresses the importance of critiquing mainstream agricultural development and fostering and nurturing alternatives and movements for alternatives as two sides of the same coin.

MODERNISATION: THE DESTRUCTION OF THE
BIOLOGICAL FOUNDATION OF AGRICULTURE

Bangladesh, with a wealth of biodiversity and natural resources, has struggled to escape poverty and underdevelopment since its independence in 1971. It has a population of over 170 million living in a small area of 147,570km. But within this small area, there is huge diversity depicted by the Agroecological Zones (AEZ) Study, which illustrates such diversity in physiography, soils, land levels above flooding and agro-climatology. It recognised 30 agroecological regions and 88 subregions, further subdivided into 535 agroecological units.[2]

Agriculture is dominated by small farm holdings (less than a hectare), which constitute 84 per cent of total farming households; only over 14 per cent are medium and large farms (over 3,000 hectares).[3] These farmers produce various crops, particularly rice. The Food and Agriculture Organisation (FAO) estimates that 500 million small family farms, owning less than a hectare of land, are the source of more than 80 per cent of the world's food supply.[4] Bangladeshi farmers belong to those categories of global farmers.

Agriculture provides livelihood and employment to most of the population and contributes to the national economy by providing food, fibre, medicine and foreign exchange. During 1983–84, the share of agriculture to gross domestic product (GDP) was 49 per cent compared to only 10 per cent for the industrial sector and 18 per cent for trade and transport. Since the 1990s, the agricultural sector's contribution to the GDP has gradually reduced from 38 per cent to only 12.9 per cent of the GDP in 2020.[5] The decline of agriculture's share in GDP is seen as a sign of 'modernisation' based on the notion that agriculture means low growth, backwardness and lack of industrialisation. Only 'industrialisation' brings high growth and civilisation. In 2020, the share of industry in the GDP had gone up to 30 per cent, and the services sector contributed about 53.4 per cent. It was an intended outcome of the policies that destroy farming as life and livelihood and turn lands into means for commercial activities, industries and industrial food production.

The fact that agriculture's contribution to GDP has declined, but continues to employ over 40 per cent of the population, does not mean much, as it fails to grasp the complex relations between people, agriculture and livelihood, particularly in agroecological zones. The shift from agriculture to so-called development and industrialisation is bringing different catastrophes in people's lives through the destruction of biodiversity, environment and health, as well as the violation of the rights of farmers and rural women.

Since the 1970s, Bangladesh had to follow the donor-driven policies of modernising agriculture called the 'Green Revolution', which is essentially the industrialisation of food production by using chemicals (fertilisers and pesticides), extraction of groundwater for irrigation and mechanisation of post-harvesting technologies. Industrial food production has been touted as a green revolution that destroyed biodiversity and promoted monoculture of high yielding variety (HYV) rice crops, gradually changing the seed technology so that farmers' seed systems could be destroyed and replaced by corporate seeds and genetically modified organisms (GMOs). Millions of tons of chemical fertilisers, pesticides and the pollution and extraction of millions of litres of groundwater created an unjust system in accessing seeds, agricultural inputs and water.

FOREGROUNDING JUST TRANSITIONS THROUGH NAYAKRISHI

In the 1990s, faced with increasing costs of inputs and lower returns on yields from conventional farming practices, farmers looked for an alternative. They were faced with the question of whether they wanted to go back to traditional agriculture or formulate a different practice that can address the problems as well as supersede modern agricultural methods and deal with new multi-dimensional emerging issues of biodiversity losses, ecological questions, farmers' and women's rights and food sovereignty. It was not a matter of going back to old times. Rather it was for future transitions. Nayakrishi Andolon was born amidst these challenges. At the outset, it was not envisioned as a technical transition from chemical-based agriculture to organic. It grasped all the social, environmental, cultural and political issues faced by farming communities.

The naming of Nayakrishi Andolon in the early 1990s was itself a challenge. The global environmental and ecological movements were active before and after the 1992 Earth Summit held in Rio de Janeiro, Brazil. The term 'biodiversity' was not familiar to many environmental activists before the Convention on Biological Diversity (CBD). Nayakrishi (Naya means *new*, Krishi means *agriculture*) evolved through discussions, debates and analysis among farmers. It was an entirely new concept of farming based on biodiversity. With it, the term Andolon (movement) was added as farmers as individuals cannot change the situation dominated by corporate interests and global players. The farmers must collectively and continuously fight against the transformation of agriculture to industrial agriculture that utilises harmful technologies such as genetically modified seeds. Nayakrishi farmers follow ten simple rules, mirroring the ten fingers of their hands. The primary aim is to maintain and regenerate living and fertile soil, maintain and regenerate diverse life forms and ecosystemic variability and develop the capacity of the indigenous knowledge system to engage and appropriate the latest advances in biological sciences that could contribute to regenerating the planet and the Earth system. These rules are routinely reviewed based on new information, practical experiences and learning. The first five rules are mandatory, such as 'absolutely no use of pesticide' or 'any chemicals and learning the art of producing soil through natural biological processes'. These are the primary obligations of being a member of the movement. Rules six to ten appeal to farmers interested in developing more integrated and complex ecological systems not only to maximise the yield but to contribute to innovating interesting ecological designs that demonstrate the immense economic potential of biodiversity-based ecological farming and strengthen the practical forms of resistance against globalisation.[6]

Resistance at the production level against chemicals and industrialisation of food production is generally known as 'organic' agriculture. However, Nayakrishi Andolon insists that food production must be based on the preservation of biodiversity, shifting from 'organic' food production to biodiversity-based agriculture. 'Organic' food production that has developed in the industrial food production system within a capitalist market, dictated by market demands, is still locked within the 'industrial', 'capitalist' and 'production' paradigm.

Agriculture is integrally related to many other livelihood occupations, such as potters, blacksmiths, weavers, fishers, livestock raisers, etc. Household is the unit of production that involves the entire family in which each member of the household plays a respective and interrelated role. In the Nayakrishi households, women become the most important contributing members of the families and are important decision-makers in their farming activities, such as seed keeping and post-harvest processing. Nayakrishi women farmers exercise their agency in the unit of production and have command over biodiversity and genetic resource preservation. Nayakrishi involves children, too, but not at the cost of depriving them of education or other social activities.

RECONSTITUTING COMMUNITY SEED NETWORK AND
KNOWLEDGE PRACTICES TO DEFEAT GLOBAL
CORPORATE SEED BUSINESS

The most effective strategy of the Nayakrishi farmers, particularly women, is the emphasis on seed preservation, collection and regeneration of the local variety of seeds. They took a community-based approach through the formation of Nayakrishi Seed Network (NSN) with the specific responsibility for ensuring both in-situ and ex-situ conservation of biodiversity and genetic resources. Farmers maintain diversity in the field but, at the same time, conserve seeds in their homes to be replanted in the coming seasons. The NSN has three levels.

First, Nayakrishi Seed Huts (NSH) are established by the independent initiative of one or two Nayakrishi farming households in the village, willing to take responsibility to ensure that all common species and varieties are replanted, regenerated and conserved by the farmers. Nayakrishi Seed Huts ensures that farmers have their own collection of seeds in their households. The diverse varieties of seeds in the farmers' households are represented in the NSH, which they can share and exchange with each other.

Second, the Specialised Women Seed Network (SWSN) is formed which consists of women having specialised knowledge in certain species or varieties. Their task is to collect local varieties from different villages. They also monitor and document the introduction of a variety in a village or locality, and keep up-to-date information about the variability of species for which they are assigned. They also watch if harmful seeds are promoted in the villages, which they can resist.

Third, Community Seed Wealth Centre (CSWC) is the apex centre in the Nayakrishi Seed Network connecting the NSH and the farmers' households. It is an institutional set-up that articulates the relationship between farmers within a village and between villages, in other districts and with national institutions for sharing and exchanging of seeds. The physical construction of CSWC is based on two principles: a) they must be built from locally available construction materials, and b) the maintenance should mirror the household seed conservation practices. These are located in one of the *Biddaghors* (learning centres) of UBINIG for seed collection, storage, preservation, distribution, exchange and regeneration. The tasks of the CSWC include documentation and maintenance of general information about the area.

Any farmer member of the Nayakrishi Andolon can collect seed from CSWC with the promise that after the harvest, they will deposit double the quantity they received. In 2021, the CSWC held a collection of over 2,700 rice varieties, and 538 varieties of vegetables, oil, lentils and spices. The CSWC also maintains a well-developed nursery with indigenous species of fruit, timber and medicinal plants.

In the CSWC, intensive interaction and sharing of knowledge and exchange of seeds are held among farmer women in each village or community, and thereby facilitates the farmers' significant progress in conserving and reproducing local planting materials. Farmers gained much confidence to continue food production through the shift to the local varieties. The farmers' seed system contributes to seed and food sovereignty in their respective communities. For Nayakrishi, food sovereignty cannot be achieved without realising seed sovereignty. They have the sovereign right to decide which food crops to grow, having the seeds in their control and not following the vagaries of the market.

Farmers of the Nayakrishi Seed Network embed these seed saving practices in their day-to-day relationships with each other and create a particular environment and agro-ecological setting to ensure their biological existence. The striking character of CSWCs and Seed Huts is their capacity to augment the dynamic and cyclical relationship between in situ and ex situ conservation of planting materials that make farming possible, sustainable and gainful. It is gainful for farmers to enhance farmers' capacity to regenerate the biological foundation of farming and generate almost all the required inputs from farming. Within the CSWCs, farmer representatives actively participate in decision-making processes.

As a movement, the primary actions of Nayakrishi Andolon are to mobilise farmers against invasive seeds such as hybrid, GMOs and any other technological aggression against the farmers' seed system. They must protect the seeds, and the slogan 'Sisters keep seeds in your hands' is central to the movement. Nayakrishi farmers have been resisting the promotion of genetically engineered crops like Bt brinjal (eggplant) and Golden Rice and have been successful in raising concerns on biosafety grounds. They also resist these technologies because they are patented by multinational companies like Monsanto (now Bayer) and Syngenta.[7]

UNCULTIVATED FOOD: REIMAGINING NATURE TO RESIST PREDATORY PRACTICES

Nayakrishi farming practices encourage the growth of uncultivated foods such as leafy greens, tubers and small fish that constitute nearly 40 per cent of the diet of the people in the communities. This is possible where local bio-diversity has been conserved. These uncultivated foods are collected from agricultural fields, water bodies and forested areas. These food sources are also important for medicinal purposes, both for people and animals. Functionally, Nayakrishi defines agriculture as the management of both cultivated and uncultivated spaces to ensure the maximum yield per acre of land – invigorating various ecological functions of the elements of living Nature.[8]

Amongst the very poor, landless members of rural communities (comprising some 15 per cent of the rural population) depend on uncultivated sources of food and fodder for nearly 100 per cent. Throughout the year, their daily survival and well-being are ensured through the collection of uncultivated foods directly and through systems of exchange with rice farmers, and the sale of goats and chickens in the local market to enable the purchase of oil and other food items they need but cannot directly forage.[9] By nurturing the growing of uncultivated food, Nayakrishi ensures the preservation of biodiversity that provides food and nutrition for the community.

The strategic role of uncultivated food and fodder in rural areas has important implications for land policies. Uncultivated food abundantly grows in common lands and spaces in rural communities. The negative consequences of privatising common areas are mainly experienced by women who rely on their surroundings for food and access to life-enhancing spaces and raw materials. Women are concerned about the privatisation of common lands and the transformation of public spaces such as roadsides and ponds, as these directly impact the livelihood options of people who depend on public spaces to graze animals or collect items for food or sale. Common areas and customary rights in these areas have been completely ignored in the policy context.

Ensuring the maintenance of uncultivated food sources in and around the immediate environment and the accessibility of common resources are necessary to realise food security within the community. The degree of control over local food sources is the measure by which government programs can ensure the capacity of poor communities to participate in the market. Rather than supplying food through state distribution systems and corporate subsidies, governments should protect and enhance locally cultivated and uncultivated biodiversity, including uncultivated food sources.[10]

RESISTING CORPORATE AGRICULTURE

Farmers are deceived by 'attractive' advertisements by companies and corporate propaganda abusing science for profit. There are hired 'experts' who claim that due to climate change, agriculture cannot be based on Nature and knowledge practices of farming communities. They understand 'Nature' as something fixed and dead, as if it is not a living being constantly changing and evolving. The companies do this to be the sole suppliers of stress-tolerant seeds. They offer so-called flood, drought and saline-resistant seeds produced in their laboratories. These seeds are patented, and farmers are compelled to buy from them at a high price. These seeds blatantly ignore the experience and knowledge of farmers in dealing with climatic variability and natural disasters! Bangladeshi farmers have many indigenous varieties: flood, drought and saline-resistant. Farmers have vast experience in adapting to changing climatic conditions for hundreds of years. Nayakrishi farmers are bringing back those seeds and also cultivating community knowledge to address climate change.

At present, one very important objective of the Nayakrishi Andolon is the protection of Planet Earth from greenhouse gas emissions and the rise of temperature, and thus, protecting the existence and the biological integrity of all life forms constituting our planet. Nayakrishi farmers persevere to keep the food chains and web of life free from hazardous industrial chemicals such as pesticides, herbicides, toxins, biocides and other harmful products, such as arsenic, that have been entering our food chains through modern industrial food production, which is one of the main causes of greenhouse gas emissions. Agriculture and the food systems are the biological foundation of civilisation. Nayakrishi farmers believe that farming is a way of life. It can save the communities by preserving biodiversity and promoting life-affirming activities.

For Nayakrishi, the new corporate terms like 'nature-based solutions' – which encompasses terms such as ecosystem-based adaptation, eco-disaster risk reduction and green infrastructure – are nothing but greenwashing. (Also see Manahan's chapter in this book.) Nayakrishi would not like to be identified as a 'nature-based solution' to the growing problem of climate change that is man-made and created by the domination of corporate control over natural resources, particularly seeds and genetic materials. Nayakrishi challenges the very idea of industrialisation and destruction of Nature as 'progress'. Through its biodiversity-based agriculture praxis, Nayakrishi works towards a just system in the community where the poor, women and indigenous people are key protagonists and political players in transforming the agri-food system of Bangladesh.

As a broad movement, Nayakrishi represents people's resistance against the destruction of the 'local' by privileging the 'global' and against effacing the 'real' by installing 'virtual'. Nayakrishi is an eco-political way to return to common sense and free farming communities from the tyranny of hierarchy, power and technology. The movement constantly explores alternative life-affirming relations and community-building practices through the critique of egocentrism, oppressive social hierarchies and the industrial notion of high-tech lifestyles and consumerism as 'development'.

CONCLUSION

Apart from practical community activities, Nayakrishi is engaged in systematic critique and resisting the idea of 'development' cast in the model of capitalist industrial civilisation. In this context, Nayakrishi is a paradigm shift from the conventional idea of development, progress and industrialisation. Nayakrishi Andolon, as a movement, imagines and nurtures ecological communities for community prosperity and joyful living. Just transition is not merely a response to livelihood and environmental disasters but a counter to the crisis and problems in our understanding of Nature and our relations and roles as human beings with it. Agriculture is an ideal site to reimagine the future of humankind where the current separation and contradiction between industry and agriculture are squarely addressed and resolved.

Nayakrishi is committed to advance the local and indigenous knowledge system and build capacities to critically integrate the success, failures and insights of formal knowledge practices such as 'modern science'. There is no mechanical separation between formal and informal knowledge systems. While Nayakrishi Andolon does not romanticise indigenous knowledge, it does not accept uncritical authority of modern science in knowledge production. Farmers are authentic knowledge producers, expressed orally or otherwise. Women farmers have demonstrated their knowledge through the practice of seed preservation and conservation, and through these, they contribute to nurturing biodiversity and all life forms.

Achieving food sovereignty is not just a slogan, it can be realised only if we care about farmers and attend to their evolving knowledge practices. The slogan of Nayakrishi women farmers 'Sisters keep seeds in your hands' calls not only for the resistance against corporate control of our food system, it ultimately demands the foregrounding of life-affirming relations between people and Nature.

NOTES

1. Climate Reality Project. 'How the climate crisis is impacting Bangladesh', www.climaterealityproject.org/blog/how-climate-crisis-impacting-bangladesh (last accessed 9 December 2021).
2. Food and Agriculture Organisation and United Nations Development Programme. 'Land Resources Appraisal of Bangladesh for Agricultural Development', Report 2: Agroecological Regions of Bangladesh.
3. Bangladesh Bureau of Statistics. 'Statistical Yearbook of Bangladesh', 2015.
4. Food and Agriculture Organisation. 'The State of Food and Agriculture (2014): Innovation in Family Farming', www.fao.org/3/a-i4040e.pdf (last accessed 9 December 2021).
5. World Bank. 'Agriculture, Forestry and Fishing Value Added. World Bank National Accounts Data, and OECD National Accounts Data Files', https://data.worldbank.org/indicator/NV.AGR.TOTL.ZS (last accessed 11 February 2023).
6. Farhad Mazhar, Farida Akhter and Upamanyu Das. 'Nayakrishi Andolon: Geolocalization Bangladesh', in *Resilience in the Face of COVID-19, Vol. 1* (Global Tapestry of Alternatives, 2021).
7. These struggles are documented by UBINIG (Policy Research for Development Alternative). For more information, see www.ubinig.org.
8. Farhad Mazhar. 'Nayakrishi Andolon', in Ashish Kothari et al. (eds), *Pluriverse: A Post-Development Dictionary* (New Delhi: Tulika Books, 2019).
9. UBINIG. 'Uncultivated food: summaries of preliminary data compiled from field reports in SANFEC, Uncultivated food that money can't buy', https://grain.org/en/article/450-uncultivated-food-food-that-money-can-t-buy (last accessed 11 February 2023).
10. Ibid.

17

Designing Systemic Regional Transitions
An Action Research Experience in Colombia

Maria Campo and Arturo Escobar

We are returning to our deepest senses, where we go back to imagining a valley of rivers and lagoons

(Vicenta Hurtado, director, Casa Cultural El Chontaduro, Cali, 13 December 2019)

INTRODUCTION: ECO-SOCIAL TRANSITIONS AS CIVILISATIONAL TRANSITIONS

In the Cauca River geographic valley in Colombia, as in many other parts of the world, eco-social, dignified and just transitions are already underway, each having greater or lesser clarity and intensity. The climate crisis and the exponential increase in inequalities linked to the extractivist model are acting as a powerful trigger for thoughts and practices aimed at weaving pluriversal bioregions centred on the care of life, proposing transitions based on just economies in harmony with the land, and building a less individualistic vision, one that guarantees the sojourn and lives of the children and youth in their territories.

This chapter is based on a regional eco-social transition trajectory and project underway within the Cauca River geographic valley in southwestern Colombia. Our project arises both from the practice of collectives in the region and from three theoretical-political concepts: pluriversal socio-ecological transitions; environmental conflicts as conflicts between worlds; and pluriversal territorial peace. This is conceived within an antiracist, antipatriarchal, anti-class, postcapitalist and territorial autonomy perspective, understood in an integrated manner. We try to move forward on the path of just transitions that are pluriversal and therefore, counterhegemonic.

The trajectory/project is being implemented by three groups:

- *Tejido de Transicionantes del valle geográfico del río Cauca* (Tapestry of Transitioners of the Geographic Valley of the Cauca River), a group of 25 people created in 2018 that includes Afro-Colombian intellectual activists, feminists, environmentalists and academics who have experience in intersectional analysis (including ableism), political ecology, transition and media design, urban–rural interfaces and human rights.

- *Asociación Casa Cultural El Chontaduro* (El Chontaduro Cultural House Association), a grassroots organisation founded in 1986 in eastern Cali, an area that constitutes one of the sectors with the city's severest social vulnerability conditions. The group is focused on building just societies with equality between genders and characterised by being ethnically and racially plural. Its lines of action – focused on work with youth, women and children, from educational perspectives, research and peacebuilding – respond to structural challenges through a commitment spurred by its Afro-feminist leadership.

- *Asociación de Consejos Comunitarios de Suarez* (Association of Community Councils of Suarez), ASOCOMS, an organisation of black communities that brings together three community councils from the area, working for the rights of the black people and of Nature through strategic actions that have included the defence of the Cauca and Ovejas rivers against energy and mining projects, the right to prior consultation on the environmental management plan of the Salvajina hydroelectric power plant, and the high-profile mobilisation of black women for the care of life and ancestral territories in 2014.

The first part of the chapter presents the regional context that makes the eco-social transition a real historical possibility. In the second, we briefly dwell on the currents of thought that serve as its inspiration, moving on to a presentation of the most relevant elements of the project in the third part. In the last section, we venture on some observations about the socio-territorial transitions and dynamics derived from our action research.

THE CAUCA RIVER GEOGRAPHIC VALLEY REGION AND THE NEED FOR AN ECO-SOCIAL TRANSITION

For centuries, the Cauca River geographic valley region has been a scenario of territorial disputes fuelled by global economic interests (agribusiness, hydroelectric, mining and energy projects), drug trafficking and armed conflict that affect food sovereignty and restrict self-determination. These issues generate and exacerbate violence against ethnicised and racialised women and girls, turning them into the spoils of war, in addition to increasing youth homicides among the black population. The government's response has been the securitisation of the territories, translated into the militarisation of the bodies and lives of women, girls and young people, affecting the spatialities historically built by ethnic peoples – all of this in the name of 'development'.

This ecological network of mountains, forests, meadows, rivers, lagoons and wetlands has been home to hundreds of species of plants and animals, as well as numerous communities and towns. This legacy has been systematically undermined by agro-industrial and extractivist operations. Touted as a miracle of 'development', the profoundly defuturising effects of this model are obvious: depletion, sedimentation and contamination of rivers and aquifers; drying up of wetlands; loss of biodiversity; deforestation and erosion of soils and slopes; respiratory problems suffered by sugarcane cutters and nearby populations due to sugarcane burning; land dispossession; forced displacement; multiple acts of violence; feminicide and youthcide; in general, the creation of a markedly unequal, anti-black, anti-indigenous and anti-peasant region. This nationally and globally articulated model is accelerating the impacts of climate change, with notable effects on all ecosystems.

One hundred years of continuous agro-industrial expansion have shown that the current development model is coming to an end. The region can easily be reimagined as a bastion of agro-ecological production of fruits, vegetables, grains and plants, organised as a pluriversal region of agricultural producers, with food sovereignty, restored territories, reclaimed soils and water sources, within a functional network of intermediate towns and cities in synergistic coexistence with the countryside – in short, an agropolitan and aquapolitan region. We define the agropolitan orientation as a perspective that synergistically integrates the countryside and the city, and that recognises indigenous and black Afro descendant peoples, and peasant communities as subjects of territorial rights with their own visions of the world and competencies to contribute to the habitability and good living of the entire population. We seek to explore the simultaneous goals of self-managed urbanisation of the countryside and selective ruralisation of the city, paying attention to the historically amphibious nature of the region. Futures of this kind are still unthinkable for the elites and middle classes, whose intensely consumerist urban-spatial way of life is inextricably linked to the globalised capitalist economy.

The Cauca River geographic valley is a scenario of ongoing transitions. Activities that can be considered transitional are embodied in multiple organisations and projects, most of them small, but increasing in number, especially relating to the following: water (pedagogy and restoration of rivers, basins and wetlands); food (peasant markets, urban gardens, land recovery, agro-ecological production and local economies); climate change (community restoration and conservation of forests, reforestation of slopes and basins); and territorial peace (indigenous, Afro descendant and human rights organisations in the region undertake important actions when it comes to the resolution of conflicts from the perspective of their autonomy). The dominant economic logic, however, goes against the current of these activities and is oriented towards the creation of a globally competitive metropolitan region, disconnected from its rurality.

The transitions underway contribute to two of the great challenges facing Colombia and the region today: effective implementation of the Final Peace Agreement signed in 2016 between the state and the FARC (Revolutionary Armed Forces of Colombia) guerrillas, and progress made in signing agreements with the National Liberation Army (ELN) and other armed groups in rural and urban areas.

CIVILISATIONAL TRANSITIONS, COSMOVISIONS/RELATIONAL ONTOLOGIES AND PLURIVERSAL TERRITORIAL PEACE

The trajectory/project is nourished by three interrelated theoretical and political fields: transitions studies; political ontology (with emphasis on relational ontologies); and territorial peace studies. In Latin America, civilisational transitions are being emphasised by a wide range of indigenous, Afro descendant, environmental, feminist and peasant movements, among others.[1] The guiding principle is the affirmation that the current crisis is a crisis of a particular mode of existence (Western, heteropatriarchal, capitalist, racist, colonial). Many transition movements are based on the re-emergence of *cosmovisiones* (cosmovisions, worldviews) or relational ontologies that reposition *the radical interdependence of everything that exists* as the true foundation of life.

The design for transitions constitutes an action research framework for reorienting localities and regions toward socially just and environmentally sustainable ways of living.[2] In Latin America, autonomous design for transitions goes hand in hand with community struggles for re-existence with an emphasis on pluriversality, understood as a world where many worlds fit.[3] Similarly, our trajectory/project is inspired by critical design studies, political ontology[4] and alternative economies.[5] We pay particular attention to relational worldviews, those that problematise the constitutive dualisms of modernity, especially the separation between the human and the non-human, and that are still present, to a greater or lesser degree, in the practices of many territorialised peoples. The principle of Ubuntu (I am because we are) embodies relational ontologies.

Colombia has been a leader in conflict and peace studies. New conceptions of territorial peace have emerged in the last decade,[6] alongside feminist approaches to peace.[7] We understand territorial peace as the socioeconomic, political, ecological and cultural framework through which a relatively stable integral balance is sought to be re-established between the multiple actors, both human and non-human, who inhabit a given territory or society. Ultimately, territorial peace has to do with the relations between worlds, or pluriversality; this is crucial for ethnic peoples and women. Our approach is intersectional;[8] making visible the network of interrelationships between economic position, ethnicity and race, gender and sexuality, age, and dis/abilities, but also territoriality (countryside and city) and cosmovision (ontology). In this way, we seek to enrich intersectionality with a spatial and ontological dimension.

The National Development Plan (2022–26) under the Gustavo Petro and Francia Márquez administration, with the motto 'Colombia, World Power of Life', constitutes a historic moment for the country, since it envisions a government policy whose priority is to safeguard life in all its manifestations, in three main areas: Total Peace, Environmental Justice and Social Justice. Black feminisms have always pursued these areas, from the mediation of conflicts appealing to *derecho propio* (autonomous conception of rights), to the enforceability of the guarantee of ethnic-territorial rights and the right to the city; but, above all, seeking to implement racial justice in Colombia, a country that functions within atavistic social hierarchies, based on racialisation, sexism and social stratification.

Our conception of a radical eco-social transition focuses on strengthening the capacity of local and regional worlds to face the socio-environmental traumas associated with social conflicts and climate change from a perspective of interdependence, intersectionality and peace. Rather than speaking of adaptation and resilience, we use the notions of re-existence and marronage, which incorporate an ecological-ontological dimension, aiming at the care and reconstitution of the fabric of life. Marronage summons forms of resistance and re-existence of the black peoples of the diaspora, prominently represented in our trajectory/project.[9]

LA TRANSICIONADA: OBJECTIVES, STRATEGIES AND ACTIONS

More than a 'project', we conceive our actions as a trajectory of a journey, which we call *La Transicionada*. In this journey, we highlight the need to imagine and craft other categories, from and for our *feel-think-act*; these imply an approach from other words and forms of communication, close to the *languaging* and *emotioning* of which Maturana and Verden-Zöller speak.[10] Playing, feeling and collaborative thinking have allowed us to create a universe of communication in our journey to build pluriversal worlds, beyond the existing hegemonies – racist, patriarchal, sexist, capitalist and ableist.

Our main objective is to contribute to a *regional socio-ecological transition* towards a socially just, environmentally sustainable, culturally diverse and pluriversal society, with territorial peace, capable of promoting food sovereignty. An essential task for this objective is to generate a new understanding and imaginary of the region that brings together transformative initiatives that are currently dispersed, calling for action under the rally cry 'another Cauca River geographic valley is possible'. This all revolves around generating conversations capable of promoting a mobilisation towards new ways of inhabiting the region and the Earth.[11] We echo the words of Father Francisco De Roux in the presentation of the Truth Commission Report, who recalls that, in order to overcome the conflict, it is through conversation that we pave the way towards the construction of anti-hegemonic counternarratives, and to reconstitute and heal the fabric of life:

> We beckon everyone to heal the physical and symbolic, multicultural and multiethnic body that we make up as citizens of this nation. (…) We beckon everyone to free our symbolic and cultural world from the traps of fear, anger, stigmatization, and mistrust. (…) We did not have to accept barbarism as natural and inevitable nor continue business and academic activity, religious worship, carnivals, and football as if nothing were happening.[12]

Along these lines, the objectives of our trajectory/project are:

1) to foster convergences among the transformative alternatives underway in the territories, particularly among those focused on the threads of water, food sovereignty, climate change, justice, peace and territorial planning, aimed at strengthening their capacity for action and coordinated strategies among them;

2) to co-design narrative and media strategies to activate the re-imagining of the region towards other possible futures through media, art, video, audiobooks, graphic design, theatre and collective authorship;

3) to promote concrete design strategies for transitions from an agropolitan perspective, initially in two locations, focused on livelihood issues, food sovereignty, climate change and territorial peace.

The objectives have been identified based on the principles of productive transformation, eco-ontological restoration (restoration of ecosystems and worlds), and historical reparations, within a pluriversal and agropolitan ethno-cultural context. These objectives should contribute to the recommunalisation of social life; the relocalisation of economic activities (such as food); the strengthening of territorial autonomy; the simultaneous depatriarchalisation and deracialisation of social relations; and the reintegration with the Earth.[13]

The actions that we are carrying out include the following: for the first objective, a) creating a map of the transformative alternatives in the region, using social and digital cartography; b) making a diagnosis of pluriversal territorial peace, focused on obstacles to peace in the territories and actions underway to overcome them; and c) building networks between transformative alternatives. The purpose is to consolidate a tapestry of transitioners capable of constituting a counterpower to the dominant narratives and initiate co-design practices towards the possibility of a different kind of Cauca River geographic valley.

For the second objective, we focus on the following actions: a) strategies for collective 'visioning' and *disoñación* ('dream-design') towards a different self-understanding of the region, using co-design tools to construct transition scenarios for the bioregion; b) multimedia production and interactive web platform that encourages conversations with broad audiences in the territories about other possible Cauca River valleys; c) proposing transition scenarios towards other possible futures and the paths to reach them. These media productions provide a space for other ways of counting, imagining, documenting, recording and designing.

The third objective is being implemented in two Afro descendant community spaces: the municipality of Suárez, in Northern Cauca, and the urban area of eastern Cali (known as the Aguablanca District), with a majority Afro-descendant population, destination and refuge for hundreds of thousands of people exiled from the Pacific and Northern Cauca. The actions are as follows: a) prepare a digital atlas of actors, actions, capacities, conflicts and knowledge for each locality, with teams of young people and women from the communities; b) workshops with communities and organisations focused on memory records and landmarks associated with spatially and temporally referenced testimonial narratives; and c) a design of strategies for environmental sustainability in the face of climate change. All of these actions are aimed at creating and putting into operation a territorial design co-laboratory through mechanisms that strengthen organisations and design opportunities that foster the transitions.

TRANSITIONS AND SOCIO-TERRITORIAL DYNAMICS

Throughout its 33 years of community process, the El Chontaduro Cultural House Association has oriented its political and cultural agency towards the territorial citizenship of the inhabitants of eastern Cali, the majority of whom are black people from the Colombian Pacific who, in search of protection and better living conditions, suffer a second victimisation within the invisible urban war they find in Cali that limits their exercise of civil liberties. Their experience shows that it is the *mujeresnegras* (black women) (triply damaged due to their race, class and gender/sex) who have generated political strategies and innovative social technologies to build territories of peace and opportunities in eastern Cali.

The processes of El Chontaduro are an expression of marronage in the present, imagined from its situated experience and with its capacity for agency based on re-existence, inhabiting a space that has been denied to black people in the city. Their praxis focuses on the construction and reconstitution of urban space from a city–countryside relation, as a diverse, happy, emotional, accomplished territory: in a word, relational, without ignoring that, even with these re-existences, the city continues to be a territory in dispute.

Art is one of the main forms of political expression in El Chontaduro. It is the way in which its people transverse the healing of intergenerational traumas, in their efforts to break the enslaving chains of past centuries; seeking to heal territories stained with the blood of the *renacientes* (the new generations) trapped by circles of violence derived from racism and inequities, where drugs and arms trafficking have found a niche. Art heals *black women* who have been survivors of the violence rooted in patriarchy. Art heals girls and boys, whose childhood is taken away by confinement during the pandemic lockdown and by the hopelessness which outside forces seek to subject this part of Cali. ASOCOMS arose as a response to the invisibilisation of the black people in the municipality of Suárez during the prior consultation process for the La Salvajina hydroelectric plant Environmental Management Plan. The plant is operated by Celsia, a subsidiary of Argos, a Colombian transnational company. Faced with the systematic violation of their rights enshrined in the Political Constitution, international conventions, and their own law that assists them as a people, the community councils of the towns of La Toma, Pureto and La Meseta in Suárez have insisted on staying united, despite actions by both the company and the government to undermine the community's integrity, sowing intra-ethnic conflict. Since 1986, when the Cauca River was flooded to build a dam, the black people and indigenous peoples of Cauca walked more than 100 kilometres as a first joint action to denounce government-led dispossession and abuses in relation to the dam project. The children, youth and adult population of those 1986 walkers came together in time to form ASOCOMS, a group that works on actions centred on agroecology, organisational strengthening, evaluation of impacts of megaprojects and gender equity, and to defend their right to exist, live in their ancestral territory and have their own options of the future.

Another emblematic fact that represents the ongoing resistance of the black people in Suárez was their opposition to the diversion of the Ovejas River to feed the dam. The people opposed a second river being taken from them, since the river represents life, given that they had already suffered the impact of the damming of the Cauca River. All these expressions of marronage mentioned herein demand freedom for communities and Nature.

A BRIEF CONCLUSION

As in other regions of the Global South, we believe in transitions that integrate calls for decarbonisation with a broad agenda of food sovereignty, post-extractivist economies, and social and pluriversal justice. Any transition process has to be oriented towards post-patriarchal, anti-racist, anti-class, agropolitan relationships and ways of life rooted in the territories. It comes down to setting up mobilisation strategies for new ways of dwelling on the Earth, capable of healing the fabric of life, with the diversity of peoples, worlds and actors that inhabit the territories of the Cauca River geographic valley bioregion. We believe that the best way to approach this task is by taking a new look at the territories from the perspective of the people who have inhabited, built, loved and suffered in them, learning from their resistance and re-existence tactics, from the experience of those who perform the tasks of care, weaving territory and community, with all those willing to contribute to reconstitute and heal the web of life.

As in other regions of the continent and the world, our journey/project constitutes a call to *a social, ecological, economic and pluriversal pact* that is capable of transitioning towards a more just and egalitarian society, based on a selective post-extractivist deglobalisation, coupled with the relocalisation of key spheres of life, which include eating, healing, learning and dwelling. This is a serious commitment to transition, in the short, medium and long term; one that is capable of facing the environmental crisis and climate collapse and healing Mother Earth, that contemplates an orderly and progressive exit from the extractivist model and substantial changes in the logic of consumerism towards pluriversal territorial peace.

NOTES

1. Arturo Escobar. *Autonomía y diseño: la realización de lo communal* (Popayán: Editorial Universidad del Cauca, 2016).
2. Ibid.; Terry Irwin. 'The emerging transition design approach'. *Design Research Society Proceedings* 3 (2018): 968–89, www.drs2018limerick.org/participation/proceedings (last accessed January 2023).
3. Ashish Kothari, Ariel Salleh, Federico Demaria, Arturo Escobar and Alberto Acosta. *Pluriverse: A Postdevelopment Dictionary* (Delhi: Tulika and Authors UpFront Publishing, 2019).
4. Mario Blaser. 'Ontological conflicts and the stories of peoples in spite of Europe: towards a conversation on political ontology'. *Current Anthropology* 54(5) (2013): 547–68; Marisol de la Cadena. *Earth Beings: Ecologies of Practice Across Andean Worlds* (Durham, NC: Duke University Press, 2015); Marisol de la Cadena and Mario Blaser (eds). *A World of Many Worlds* (Durham, NC: Duke University Press, 2018).
5. J.K. Gibson-Graham, Jenny Cameron and Stephen Healy. *Take Back the Economy. An Ethical Guide for Transforming our Communities* (Minneapolis, MN: University of Minnesota Press, 2013); Natalia Quiroga, *Economía pospatriarcal* (Buenos Aires: Lavaca, 2019).
6. Astrid Ulloa and Sergio Coronado (eds). *Extractivismos y posconflicto en Colombia: retos para la paz territorial* (Bogotá: CINEP/Universidad Nacional, 2016).

7. Diana M. Gómez. 'De la inclusión de género a la estructuración de la paz feminista: Aportes de los feminismos decoloniales al proceso transicional en Colombia', in *Comisiones de la verdad y género en el Sur Global: miradas decoloniales, retrospectivas y prospectivas de la justicia transicional* (Bogotá: CIDER, 2021).

8. Mara Viveros Vigoya. 'La interseccionalidad: una aproximación situada a la dominación'. *Debate Feminista* 52 (2016), http://dx.doi.org/10.1016/j.df.2016.09.005 (last accessed October 2022).

9. Vicenta Moreno. 'Ay Dios baja y ve cómo las mujeres afrocolombianas resisten al destierro'. *Revista CS* 12 (2012): 415–34, http://www.scielo.org.co/pdf/recs/n12/n12a13.pdf (last accessed January 2023); María Campo y *otras negras ... y feministas!* 'Los feminismos, si son tales, tienen la tarea de ir a las raíces de las opresiones'. Presented at the conference, Corpus Africana: Danser et Penser L'Afrique et ses Diasporas. Toulouse; Elba Mercedes Palacios. 'Sentipensar la paz en Colombia. Oyendo las re-existentes voces Pacíficas de mujeres negras afrodescendientes'. *Memorias: Revista Digital de Historia y Arqueología desde el Caribe colombiano* 38 (2019): 131–61; Colectivos Otras negras... y ¡feministas!, Elba Palacios, María Mercedes Campo, Martha Rivas, Natalia Ocoró and Betty Ruth Lozano (eds), *Feminicidios y Acumulación global. Memorias del Foro Internacional sobre Feminicidios* (Quito: Editorial Abya Yala, 2019).

10. Humberto Maturana and Gerda Verden-Zöller. *Amor y Fuego: fundamentos olvidos de lo humano desde el patriarcado a la democracia* (Santiago de Chile: JC Sáez, 1993).

11. There is a phrase widely used by elder women in the Colombian Pacific Littoral: 'In the Pacific we don't speak, in the Pacific we sing', and when we talk about conversing, we are referring to the possibility that words have a rhythm like the drum that reproduces the beating of the heart and act to heal us, that we speak with our whole body and go through our emotions to think collectively.

12. Comisión de la Verdad de Colombia. *Hay futuro si hay verdad: Informe Final de la Comisión para el Esclarecimiento de la Verdad, la Convivencia y la No Repetición* (Bogotá: Comisión de la Verdad, 2022).

13. Arturo Escobar. 'Reframing civilization(s): from critique to transitions.' *Globalizations*, 2022, DOI: 10.1080/14747731.2021.2002673; Spanish version: ARQ (Santiago de Chile), 112, http://dx.doi.org/10.4067/S0717-69962022000 200024 (last accessed January 2023).

18

Towards a New Eco-Territorial Internationalism

Breno Bringel and Sabrina Fernandes

INTRODUCTION

The internationalist call has echoed across social movements and workers' organisations ever since the global ecological crises and the interconnectedness of experiences under colonisation and capitalism became more evident. The anti-capitalist and workers' movement has a long history of internationalist organising, perhaps better known in the past through the formation, dissolution and re-formation of various communist Internationals. After and beyond the Four Internationals, internationalisms (in plural) have become much more complex over recent decades. The emergence of transnational networks and coalitions (many of which are central to the advocacy agenda), the construction of global spaces of convergence (such as the World Social Forum), the formation of the global justice movement (with multiple expressions of the struggle against capitalist globalisation) or the internationalisation of territorialised movements, such as indigenous and peasant ones (Via Campesina is a well-known case) are only some examples of the diversity of recent internationalist articulations.

The worsening of a global ecological crisis in the twenty-first century demands an internationalist articulation which puts Nature at the centre and establishes that no popular and anti-capitalist movement can triumph and survive in an alternative society without securing the ecological conditions for life, especially a dignified life. This internationalism needs to critique global asymmetries and take an anti-imperialist stance, challenging the ties between the international division of labour, green colonialism and ecological imperialism in its thirst for resources and the continuous generation of sacrifice zones. Whereas ecological imperialism plunders Nature into resources that are continuously fed from peripherical territories into the industrial centres of capitalism, green colonialism is about how old practices of appropriation and dispossession now take on a 'green' façade by taking control of key elements of the ecological transition such as minerals for electric vehicle batteries or hectares of forest for carbon credits. These processes further exacerbate ecological debt and the related North–South asymmetries, which must be addressed without neglecting the importance of alliances between the Global South and the North. Instead of focusing only on change from above or on campaigning, we need an eco-territorial approach to contemporary internationalisms. This does not negate the necessity to struggle at the different state levels but emphasises the need to better articulate territorial conflicts across different regions and continents to avoid both localist or only macro approaches to the multiple crises of our time.

This chapter discusses the contemporary ecological crisis and provides a brief outlook into the significant transformations of internationalisms and the struggles for global justice over the last three decades, drawing attention to the growing articulation between climate and territorial conflicts. As a result, both a diagnosis and a navigation compass are proposed for transformative eco-social struggles in the contemporary world.

POLYCRISIS, GREEN COLONIALISM AND NEW ECO-TERRITORIAL CONFLICTS

Our time presents multiple crises that impact diverse parts of the globe differently. However, the analysis' layers and root causes vary according to the direct interests at play. From a just transition perspective, climate change is not just one more crisis, but an emergency that also adds to the severity in degree and scale of other challenges. It is part of the polycrisis we face, which is not simply the sum and combination of multiple crises, but the emergence of a widespread phenomenon that results from the tensions, ambiguities and contradictions of the crises in their contribution to systemic risks. Scott Janzwood and Thomas Homer-Dixon argue that, in a polycrisis, a 'temporal alignment of systemic risks' can produce 'synchronous failure of the interconnected systems'.[1]

This is of particular interest to those working on socio-ecological transitions because of the inequalities surrounding systemic risks once we consider history, means and ends. Past and present colonial relations have created a scenario where communities and nations are expected to handle the impacts of climate change and take the measures necessary to mitigate and adapt to it, but there are conflicts over the resources needed to transition and over the exact destination of our society's transition. A society free of ecological crisis does not look the same to everyone, since political projects and their end goals reflect dominant perspectives on the mode of living. There is a big difference between a project of ecological transition that requires moving away from the current extractivist model and one that preaches a green and sustainable society in one part of the world through the creation of sacrifice zones elsewhere.

Because climate change creates a systemic risk that alters and amplifies other systemic risks at several scales, a transition is necessary, even if it means different things to different actors. Nowadays, the idea of a just transition, once tied to job guarantees and redressing asymmetries in terms of who caused the crisis and who suffers its worst consequences, has slowly been appropriated by corporations, who have watered it down to pricing schemes as they become more and more present in official climate negotiation spaces. Rather than abandoning the notion of just transition, we need to define the criteria for this framework, recover its roots in labour organising and environmental justice, and place it within the many layers of decision-making and political arrangements, from a small community to national borders and international agreements. Just transition must create systemic patterns of resilience that address political, economic, social and ecological concerns, recognising that different actors in the polycrisis may collaborate but also compete for resources and priority.

The challenge posed by the new phase of extractivism associated with green transition economies helps to illustrate this (see the article by Kristina Dietz in this book). Green extractivism is the name given to new extractive ventures that disrupt ecosystems and communities in order to provide resources to companies and countries for climate change mitigation and adaptation policies. As we saw in Svampa's chapter in this book, one of its most visible faces is the battle over lithium and its value in energy transition projects and the electrification of transportation. The International Energy Agency is already predicting a lithium shortage by 2025, given its estimates of how many electric vehicles should be on the road to replace conventional ones and reduce road transport emissions to Net Zero targets.[2] The problem is that the solution presented to a crisis – vehicle electrification as the answer to transportation emissions – exacerbates traditional patterns of ecological imperialism and its unequal ecological exchange, increases political-economic pressure on dependent economies, and greenwashes the creation of new sacrifice zones in areas of high extractivist interest for newer green colonial practices. More affluent societies and the corporate interests that dominate them prefer to normalise more extractivism in the name of technological substitution rather than restructuring demand and production in ways that secure a horizon for multiple transitions – at the energy level, away from climate change and towards a post-capitalist ecological society – with eco-social transformations at home *and* in more vulnerable and poorer places.

Green extractivism does not address the root cause because it isolates the need for decarbonisation – a primary demand for energy and climate transition – from the metabolism of Nature. A simple substitution of today's hegemonic, consumption-intensive mode of living with a carbon-free alternative is materially impossible given finite resources, but also ecologically and socially undesirable given the impacts on ecosystems, communities and territories. In other words, green extractivism normalises this imperial mode of living,[3] understood as a process of hegemony and subjectivation that promotes production and consumption practices incompatible with the metabolism of Nature for certain minority groups and places at the expense of others. It is the kind of response to the climate crisis that leaves capitalism unquestioned and contributes to a problem of feedback loops and cross purposes,[4] where a false solution exacerbates the systemic risks that solutions are supposed to address.

The maintenance and creation of new sacrifice zones add to the bulk of territorial conflicts already worsening due to geopolitical tensions and wars, the short- and long-term effects of extreme weather events and the loss of habitat and livelihoods, which are linked to a larger migrant and refugee crisis. It is difficult to define what constitutes a climate migrant/refugee, especially in the context of the polycrisis, as mobility decisions cannot be reduced to a single factor.[5] Droughts, hurricanes, floods, heat waves and rising sea levels will affect people differently according to class, race, gender, location, etc. A concept of migration justice that includes both the right to leave and the right to stay, helps to highlight changing patterns and issues of access to resources, closed borders and the increasing securitisation of migration in the context of the Anthropocene. Again, systemic risks are compounded when mobility is tied to capital, which determines direction, security and means of livelihood as territories are severely impacted by climate change. It also means that internationalism is still needed today to link gains and losses to territories facing neocolonial practices.

Moreover, authoritarian trends point to the risks associated with violence and territorial claims based on various justifications. While traditional colonial practices of dispossession have been met with long-term resistance, including those against fossil fuel capital and its industrial extractivist practices, the threat of eco-fascism is also emerging to give authoritarianism a 'green rationale'. Eco-fascism normalises social and economic inequalities and penalises the poor and racialised peoples to secure territories and state borders for those who are seen as the proper stewards of ecosystems. It also finds support in green extractivism, ensuring that sacrifice zones remain far from the territories claimed by the far right, maintaining capitalism as an economic system, seeking a green version of an imperial mode of living and reproducing global inequalities and segregation. The reality of multiple threats and additions to systemic risks thus calls for careful consideration of the role of left internationalism in tackling the crises from multiple poles and scales.

FROM 'ANOTHER WORLD IS POSSIBLE' TO 'ANOTHER END OF THE WORLD IS POSSIBLE'

After the fall of the Berlin Wall, internationalist praxis was deeply restructured due to the reconfiguration of social actors and the societal and geopolitical changes that led to an unprecedented globalisation of processes, structures and flows. In this new scenario, the nation-state was decentred as the hegemonic reference for protests and the articulation of social actors, and interactions between scales became more dynamic. Since then, we have witnessed a 'progressive denationalisation of internationalism'.[6] If in the mid-twentieth century internationalism was articulated around solidarity with revolutionary projects in different states (Cuba in Latin America, Vietnam in Asia, or the various African states during decolonisation), at the end of the twentieth century it began to be built around territorialised experiences and concrete social movements, as in the case of the EZLN (Zapatista Army of National Liberation) in Chiapas or the MST (Landless Workers' Movement) in Brazil.

Internationalist praxis also became more fluid, including international organisations and their members, and waves of uprisings and the organisation of mobilisation calendars and spaces of convergence. Moments that connect crises in different regions of the world tend to trigger these waves, as could be the case with May 1968 and the anti-war movement of that time. The trajectory of these waves in the first two decades of the twenty-first century helps to identify thematic shifts in how people and movements converge internationally, and to show how the transversal nature of the ecological crisis adds threats and intensifies concerns, thus shaping the approach to nations and territories and the question of alternatives. We have identified three main moments in this new century that coincide with critical junctures where global crises and new horizons for internationalist action have occurred simultaneously.

Struggles for another world

The first moment coincides with the beginning of the new millennium and is marked by two critical events in 2001. On the one hand, the 9/11 attacks in the United States and the subsequent war in Afghanistan led to an upsurge in global securitisation and militarisation. On the other hand, in the same year, the first edition of the World Social Forum (WSF) was held in Porto Alegre to create a space of confluence to discuss and deepen proposals, exchange experiences and articulate social movements and different networks against neoliberalism and imperialism.

The WSF was the essential propositional part of the alter-globalisation movement, which has been organising protests against the main symbols of neoliberal globalisation since the early 1990s. To this end, it created internationalist coordination mechanisms such as the Global Peoples' Action and consolidated global interpretative frameworks, identifying common enemies for the struggles on all continents (transnational corporations, international organisations, multilateral agencies, etc.). The alter-globalisation movement has carried out different types of actions: counter-summits against the international financial institutions, summits parallel to the official ones (in some cases, challenging the official agendas; in others, building their own agendas), and spaces of convergence for global protests and proposals. More than that, it has consolidated an alter-activist culture,[7] critical of capitalist power and economic growth, with horizontal practices that drew on the legacy of the post-1968 movements and were inspired by the Zapatistas. Participants were aware of the importance of combining individual action with systemic critique and local action with global intervention.

The environmental movement actively integrated itself into the alter-globalisation movement and contributed to its success. Environmental justice was gradually integrated into the global justice movement. Demands for the cancellation of foreign debt, the recognition of the ecological debt of the North and the struggle against free trade went hand in hand with critiques of development, Eurocentrism, colonialism and patriarchy. At the same time, the protagonism of the indigenous, peasant and grassroots movements has enabled new formulations of alternatives based on territorial conflicts, such as food sovereignty and water justice.

The thematic and regionalisation of the World Social Forum extended it to many regions of the world and allowed it to adapt to more specific areas of action, such as education, migration and many others. In the end, however, the main signs of the movement's weakening became visible. Its trial by fire was the economic and financial crisis of 2008, when, having exposed capitalist globalisation and its social, economic and environmental consequences, the movement failed to articulate a global response when a new critical juncture appeared.

Although the 2007/08 crisis was initially presented as a subprime mortgage crisis, its scope was quickly expanded to a broader crisis of the global financial system, articulated with several other crises, especially those related to food and the environment. Despite the diagnosis of a multidimensional and systemic crisis, most social struggles devoted their energies to confronting, in a more reactive way, the immediate consequences, such as austerity measures, in their own countries. The decline of unified action at the international level against global structures, and the political and discursive re-appropriation of many of the demands of the alter-globalisation movement by hegemonic forces then led to a new cycle of mobilisations around the world from 2010 onwards: the protests of indignation.

Square protests around the world

While this new cycle was global in scope, its mobilisations took place at the national level, creating fluid connections between struggles in a kind of 'internationalism of resonances'.[8] The struggle for democracy, social justice and dignity were common points that took on diverse meanings and specific demands in different cases, such as the Arab Spring, the Indignant Movement and Occupy Wall Street.[9] These 'square movements' were articulated internationally but differently from the alter-globalisation movement. Thanks to new technological possibilities of Facebook, Twitter and other digital social media, their repertoires, messages, worldviews and demands travelled much more quickly and virally around the world, being redefined and adapted with astonishing ease. Yet no permanent transnational spaces were built to allow a deeper understanding of the struggles, subjectivities and realities of other places. Global days of action were called, but their international diffusion was produced by mobilising local nodes without a solid international articulation.

The rejection of the hegemonic political systems, traditional political parties and conventional forms of political organisation was transversal to all expressions of this global cycle of anger. More than a critique of capitalism per se, the consensus point was the fatigue of traditional politics. However, the window of opportunity opened by these protests did not necessarily lead to the democratisation of societies and political systems. On the contrary, capitalism became more authoritarian, and far-right forces proliferated, even building their international articulations, in what Tamayo calls an 'international of hate'.[10] Inequalities increased, as did the North–South divide. Many social and political forces became hostages to short-sighted policies that prevented them from looking beyond their navels, and to political polarisations that reduced the complexity of the world.

Meanwhile, the second half of the 2010 decade saw an increase in climate strikes. Youth activism began to draw media attention to the climate emergency, especially in the Global North. At the same time, land defenders in the Global South gained prominence by combining immediate resistance with the everyday care of our 'common home'.

The pandemic and the emergence of a new global moment for internationalist struggles

Despite the many warnings about systemic risks, few imagined that a virus would cripple the world in 2020. The COVID-19 pandemic came at a historic moment of natural resource depletion and climate and environmental emergency, when capitalism was at its most predatory stage. It was a time of setbacks for democracy and human rights, of distrust and rejection of politicians. Public services had been dismantled by decades of neoliberalism, which had penetrated far beyond the economy into individual and collective subjectivities. Meanwhile, the digitalisation of society had enabled a greater flow of information about the pandemic. However, this was accompanied, before and after COVID-19, by a process of growing individualisation, the spread of fake news, and increased surveillance and social control based on shared data and corporate ownership of digital tools. Thus, the coronavirus pandemic opens a third critical juncture that places us in a still-evolving scenario of future struggles.[11]

Three different projects are currently competing for the direction of the post-pandemic world. The first one is that of business as usual, focused on GDP growth, predatory developmentalism and the search for new market niches to pull economies out of the crisis, including adjustment policies that, once again, require sacrifices by the majority to maximise profits for the few. Second, green capitalism is usually associated with hegemonic ecological modernisation strategies (led by corporations and most states), primarily concerned with what is understood as the energy transition (the shift to renewable energies and the necessary infrastructure), with the private sector as the driving force. And finally, the paradigm shifts towards a new environmental and socio-economic matrix, proposed by various actors, ranging from scientific communities and progressive religious movements to grassroots actors and many anti-capitalist sectors that see degrowth, Buen Vivir ('good living'), eco-socialism and more disruptive measures as the only possible alternative to neoliberal capitalism. These projects seem to open up three possible scenarios, which do not occur in a 'pure' mode and can be intertwined in multiple ways, although all have their own logic: the recovery of the most aggressive face of economic growth; the adaptation of capitalism to a 'cleaner' model for some parts of the world, although socially unequal and still based on predatory relations with Nature, especially in the geopolitical South; or an eco-social transformation to a new model, which implies multiple transitions to build radical changes in the ecological, social and economic matrix as well as in international relations.[12]

Therefore, a new wave of internationalism for our historical moment must be an internationalism that articulates and promotes globally just transitions as an aggregating element that allows the denunciation of green capitalism and the building of radical horizons of transformation. We must start from the lessons of previous attempts at social change, while recognising the diversity of existing territorialised initiatives that already are sowing the seeds of a better world in their daily lives. 'Other possible worlds' which, today, no longer share the optimism characteristic of the alter-globalisation movement, but are rather marked by a clear awareness of finitude. This is why we are moving from 'another world is possible' to 'another end of the world is possible', which means living closer to the earth, and territorially addressing the different faces and temporalities of the collapse. At the same time, we need to build global horizons of common transformation, based on ecological justice and the concrete needs of people. This is what we call *eco-territorial internationalism*.

TOWARDS AN ECO-TERRITORIAL INTERNATIONALISM

Eco-territorial internationalism is a social practice and a form of transnational articulation between experiences linked by the impact of socio-environmental conflicts and by the construction of concrete territorial alternatives of just transitions in different areas, such as energy (community, decentralised and democratic energies), food (agro-ecology and food sover- eignty), production (workers' control of production sites) and consumption (relocation and solidarity economy), life care or infrastructure and collective mobility (dignified and efficient housing, and sustainable ways of moving, living and socialising in the city). They are localised experiences, but not strictly local because they have acquired what Doreen Massey has called a 'global sense of place'.[13] As Martin Arboleda has argued, for example, ter- ritorial struggles against extraction help to show that while the experience under capital is fragmented, the effects of certain productive practices lead to simultaneous class experiences that reveal the interconnectedness of what is produced, how, what for, where and for whom.[14]

It is a new kind of emerging internationalism that articulates the trans- versality of the environmental question with what Svampa called the eco-territorial turn of social struggles.[15] It should recover several legacies of the internationalist experiences mentioned above and articulate other horizons of eco-social transformation. For example, the alter-globalisation movement has taught us how to organise strong and coordinated global actions but with territorial autonomy. Denouncing the main capitalist actors and the global structure (and not only the impacts of their actions) is some- thing to be saved. On the other hand, the movement of the squares, despite its many contradictions, has shown us the importance of extending the space of struggle to sectors of the population that are not usually mobil- ised. It has also taught us positive lessons about digitalisation (with multiple possibilities for training and diffusion), although in the end it relied too much on corporate digital networks instead of creating its own communica- tion networks – something that has characterised the counter-information dynamic of social movements for global justice since the Zapatistas.

In terms of spatialities, eco-territorial internationalism presupposes a critique of an anthropocentric conception of scale, which, although dynamic, is usually limited to formal scales understood as levels of action (local, regional, national and international). As an alternative, a relational and bio- centric politics of scale has emerged, which considers the body as the first political scale in its multiple relations with territories.[16] Nature itself appears as a fundamental element of this new politics of scales, since the articulation between struggles and experiences of just transitions can be displaced from the 'local' or the 'national' to be located, for example, at the level of a bioregion or a specific watershed. In this perspective, rivers, forests and biomes are essential for the construction of resistance and broader alternatives.[17]

Eco-territorial internationalism can be constructed in different ways. In some cases, it can emerge from existing formal or informal networks that bring together communities and organisations from the frontlines of resistance that are currently struggling to build truly just transitions in their communities. This is the case, for example, of the Climate Justice Alliance–Communities United for a Just Transition, which brings together more than eighty organisations from different regions of the United States. Although it is a national platform, it recognised that just transitions cannot be built only in one place or country. There is, therefore, a predisposition to participate in broader internationalist initiatives based on territorialised struggles. In other cases, regional and internationalist platforms have already formed or under formation. In Latin America, for example, the Eco-social and Intercultural Pact of the South is a recent initiative that seeks to give visibility to experiences of socio-ecological transformation in the region, and to build proposals, new political imaginaries and bridges with experiences from the North and other parts of the Global South. An initiative that integrates struggles and projects of transition in Latin America is important because, in terms of political economy, the region is a target for extractivist activities that create sacrifice zones to extract materials and produce goods to maintain the imperial mode of living elsewhere. An internationalist regional outlook strengthens against the new green colonial perspectives being imported into the region, for example on who gets to exploit and export the most lithium stocked in Chile, Argentina, Bolivia (and more recently Peru), needed for the global energy transition, by denouncing how, in reality, the race for lithium does not currently follow global justice frameworks around energy democracy, but rather feeds the energy transition of the Global North and its energy-intensive demands.[18] Shifting scales, the Global Tapestry of Alternatives, as a global network of networks, seeks to build bridges between alternatives around the world and promote new spaces of collaboration, exchange and confluence. What is distinctive about these spaces is that they are committed to building alternatives to development and radical systemic transformations – or eco-social transformations – as a horizon. In the short term, however, the Just Transitions paradigm emerges as an aggregating framework for urgently needed policies that confront the contradictions of changing course while articulating concrete initiatives that already prefigure the world we would like to have.

The importance of articulating internationally the experiences of just transitions in different parts of the world is to build a common platform of struggles, pointing to common actions (such as caravans, marches, occupations, infrastructure disruptions and global actions) and different horizons of eco-social transformation. However, it is also crucial to identify common enemies and share challenges and good practices while respecting diversity and contextual specificities. Protecting the framework of a just transition from green capitalism is also an urgent challenge. Changes and opportunities also include sharing what has worked and what has not; identifying new ways and synergies for building resilient communities that resist capitalism and begin to move towards post-capitalist horizons; working on common forms of pressure and advocacy; creating common concepts and political flags and planning production across borders. This last point counters today's global patterns of production dictated by trade balance, comparative advantage, multinational corporations and free trade agreements, and instead works towards a planned allocation of resources according to the most important needs for a good life.

If anti-colonialism and anti-imperialism were a historical vector of internationalism, this is still valid, although loaded with new meanings. The struggle against ecological imperialism and green colonialism, materially expressed in green extractivism, is now a key issue. It must be accompanied by the decolonisation and depatriarchalisation of everyday relations and subjectivities. Anti-imperialist stances derived from old notions of national sovereignty, such as in the case of Global South countries arguing for their right to exploit their own oil and gas, are based on a now obsolete understanding of sovereignty.[19] Anti-imperialism in the polycrisis requires ecological sovereignty as a principle, which calls for autonomy, cooperation, solidarity and political action across borders.

Therefore, a new eco-territorial internationalism is essential to build common ground and a new synthesis towards a horizon of eco-social transformation, aggregating and articulating struggles against green capitalism and around just transitions, as plural as they may be. In practice, movements here and there are already articulating with each other in calls for cooperation and solidarity, but eco-territorial internationalism still needs to be structured and enlarged to strengthen a global, intersectional and decentralised movement of movements that could bring together actors and movements working on transformations around housing, food, societal–Nature relations, energy, race and class in an unprecedented way.

NOTES

1. Scott Janzwood and Thomas Homer-Dixon. 'What is a global polycrisis?', Cascade Institute, 16 September 2022, 5, https://cascadeinstitute.org/technical-paper/what-is-a-global-polycrisis/.

2. International Energy Agency. 'The role of critical minerals in clean energy transitions'. World Energy Outlook Special Report, last modified March 2022, https://iea.blob.core.windows.net/assets/ffd2a83b-8c30-4e9d-980a-52b6d9a8 6fdc/TheRoleofCriticalMineralsinCleanEnergyTransitions.pdf; 'Lithium supply is set to triple by 2025. Will it be enough?', S&P Global, 24 October 2019, www. spglobal.com/en/research-insights/articles/lithium-supply-is-set-to-triple-by-2025-will-it-be-enough.

3. Ulrich Brand and Markus Wissen. *The Imperial Mode of Living: Everyday Life and the Ecological Crisis of Capitalism* (London: Verso Books, 2021).

4. Adam Tooze. 'Chartbook #165: Polycrisis – thinking on the tightrope', *Chartbook*, 29 October 2022, https://adamtooze.substack.com/p/chartbook-165-polycrisis-thinking.

5. Andrew Baldwin, Christiane Fröhlich and Delf Rothe. 'From climate migration to Anthropocene mobilities: shifting the debate'. *Mobilities* 14(3) (2019): 293.

6. Breno Bringel. 'Social movements and contemporary modernity: internationalism and patterns of global contestation', in Breno Bringel and José Mauricio Domingues (eds), *Global Modernity and Social Contestation* (London: SAGE, 2015), pp. 122–38.

7. Jeffrey Juris and Geoffrey Pleyers. 'Alter-activism: emerging cultures of participation among young global justice activists'. *Journal of Youth Studies* 12 (2009): 57–75.

8. Paolo Gerbaudo. 'Protest diffusion and cultural resonance in the 2011 protest wave'. *The International Spectator* 48(4) (2013): 86–111.

9. Marlies Glasius and Geoffrey Pleyers. 'The global moment of 2011: democracy, social justice and dignity'. *Development and Change* 44(3) (2013): 547–67.

10. Juan José Tamayo. *La internacional del odio* (Barcelona: Icaria, 2020).

11. Breno Bringel and Geoffrey Pleyers (eds). *Social Movements and Politics during Covid-19* (Bristol: Bristol University Press, 2021).

12. Breno Bringel. 'Covid-19 and the new global chaos'. *Interface: A Journal for and about Social Movements* 12(1) (2020): 392–99.

13. Doreen Massey. 'A global sense of place'. *Marxism Today* (1991): 24–29.

14. Martín Arboleda. *Planetary Mine: Territories of Extraction under Late Capitalism* (London: Verso Books, 2020).

15. Maristella Svampa. *Neo-extractivism in Latin America* (Cambridge: Cambridge University Press, 2019).

16. Astrid Ulloa. 'Repolitizar la vida, defender los cuerpos-territorios y colectivizar las acciones desde los feminismos indígenas'. *Ecología Política* 61(2021): 38–48.

17. Arturo Escobar. *Territories of Difference: Place, Movements, Life, Redes* (Durham, NC: Duke University Press, 2008).

18. 'Manifesto for an Eco-social Energy Transition from the Peoples of the South', Eco-social and Intercultural Pact of the South, 2023, https://pactoecosocialdelsur. com/manifiesto-de-los-pueblos-del-sur-por-una-transicion-energetica-justa-y-popular-2/.

19. Sabrina Fernandes. 'Sovereignty and the polycrisis', in *The War in Ukraine and the Question of Internationalism* (London: Alameda Institute, 2023), https://alameda.institute/2023/04/30/xi-sovereignty-and-the-polycrisis/.

Notes on Contributors

Alberto Acosta is an Ecuadorian economist and companion to many social movements and struggles. He is a Professor Emeritus at FLACSO-Ecuador (Latin American Faculty of Social Sciences). He served as Ecuador's minister of energy and mining in 2007 and presided over the Constituent Assembly (2007–08) which crafted the Constitution that introduced the rights of Nature. Among numerous publications, he co-edited the book *Pluriverse – a Post-Development Dictionary* in 2019.

Bengi Akbulut is an associate professor at Concordia University, Montréal. Her research is in political economy, ecological and feminist economics. She has written extensively on the political economy of development, with a particular focus on Turkey. Her recent work focuses on economic alternatives, including community economies, economic democracy and degrowth.

Farida Akhter is one of the founders of UBINIG, a Bangladeshi NGO which has set up one of the biggest community seed banks in the world. She has carried out extensive research in agriculture, marine fisheries and development. She is deeply involved in the farmer-led movement Nayakrishi Andolon (New Agriculture Movement), which advances biodiversity-based farming systems involving over 300,000 farming families in the country.

Nnimmo Bassey is an architect, environmental activist, author and poet. He chaired Friends of the Earth International from 2008 to 2012 and is the executive director of Health of Mother Earth Foundation in Nigeria. He also serves on the Steering Committee of OilWatch International. His books include *To Cook a Continent: Destructive Extraction and the Climate Crisis in Africa*.

Pablo Bertinat is an electrical engineer from Argentina, professor at the Universidad Tecnológica Nacional, Director of the Energy and Sustainability Observatory of the Universidad Tecnológica Nacional, and member of the Taller Ecologista de Rosario. He works on energy transition within the framework of socio-ecological transformation processes. He is a collaborator of several networks and social movements, and co-coordinates the Energy and Equity Working Group.

Ulrich Brand is a professor and researcher at the University of Vienna, Austria. His work focuses on international politics and political economy, state transformation and its internationalisation, imperial mode of living, social-ecological transformation, international environmental and resource politics. He recently co-authored the book *The Imperial Mode of Living* (Verso, 2021).

Breno Bringel is a Brazilian activist-scholar. Professor of Sociology at the State University of Rio de Janeiro and Senior Fellow at the University Complutense de Madrid. He is editor of *Global Dialogue* and member of the Eco-social and Intercultural Pact of the South. His recent research deals with the transformation of social movements, and the geopolitics of eco-social transitions.

Maria Campo is a Colombian black feminist. She is a member of the Asociación Casa Cultural El Chontaduro and the Tejido de Transicionantes por el Valle Geográfico del río Cauca. She is involved in the defence of ethnic-territorial and black women's rights. She studied philosophy at Univalle and has a master's degree in regional development and territorial planning at the Universidad Autónoma de Manizales.

Christian Dorninger works as an interdisciplinary researcher at the Institute of Social Ecology, University of Natural Resources and Life Sciences (BOKU) in Vienna. His research examines the global distribution of environmental goods and burdens. His recently published articles explore the effects of ecologically unequal exchange on development aspirations.

Kristina Dietz is Professor of International Relations with a focus on Latin America at the University of Kassel, Germany. Her main study fields are political ecology, political economy, critical agrarian studies, social movement studies and democracy studies. Her most recent book is the *Handbook of Critical Agrarian Studies* (together with H. Akram-Lodhi, B. Engels and B. McKay, 2021).

Arturo Escobar is an activist-researcher from Colombia and a professor of anthropology emeritus at the University of North Carolina, Chapel Hill. Over the past 30 years, he has worked closely with Afro-descendant, environmental and feminist organisations in Colombia. His most recent book is *Pluriversal Politics: The Real and the Possible* (2020).

John Feffer is the director of Foreign Policy in Focus at the Washington-based Institute for Policy Studies. He is the author of several books, including *Aftershock: A Journey into Eastern Europe's Broken Dreams*, and his articles have appeared in *The New York Times, Washington Post, USA Today* and many other publications.

Sabrina Fernandes is a Brazilian sociologist, writer and eco-socialist activist. She was a postdoctoral fellow focused on Latin America and the Anthropocene at CALAS in Guadalajara and is currently a senior research advisor to the Alameda Institute. Her current research draws from Marxist ecology to discuss just transitions and their contradictions. She was formerly an editor at Jacobin and a fellow with the Rosa Luxemburg Stiftung IRGA.

Luis Gonzalez Reyes is a Spanish activist specialising on environmentalism, pedagogy and economy with a PhD in chemical sciences. He is a member of Ecologistas en Acción, part of Garúa cooperative and responsible for eco-social education in FUHEM. He is the author of several books.

Hamza Hamouchene is an Algerian researcher-activist and a founding member of the North African Food Sovereignty Network. He is a prolific writer who has written extensively on energy extractivism, democracy, anti-colonial struggles and climate justice in North Africa. He is currently the Coordinator for the North Africa program of Transnational Institute.

Rachmi Hertanti is an Indonesian lawyer and activist. She has been involved in several just trade policy advocacy and campaigns, focusing on investment treaties and trade-related energy raw materials. She was the former Executive Director of Indonesia for Global Justice and currently works as a researcher with Transnational Institute (TNI).

Edgardo Lander is a Venezuelan sociologist and is considered a leading thinker on the left in Venezuela, where he is a retired professor at the Central University. He teaches at the Universidad Andina Simón Bolívar in Quito and the Universidad Indígena de Venezuela. His recent research focuses on civilisational crisis and coloniality of knowledge, progressive governments in Latin America, as well as the influence of the new right on environmental policies in the Americas.

Miriam Lang is an activist academic and Professor of Environment and Sustainability at Universidad Andina Simón Bolívar, Ecuador. Her research focuses on development critique and systemic alternatives. She combines decolonial and feminist perspectives with political economy and political ecology. She is part of the Eco-social and Intercultural Pact of the South.

Mary Ann Manahan is a Filipina feminist activist researcher and currently, a doctoral assistant with the Conflict Research Group of the Department of Conflict and Development Studies at Ghent University in Belgium. Her current research focuses on the intersections of indigenous peoples' self-determination, forest conservation and alternatives to development. She has co-coordinated the Beyond Development Global Working Group since 2020.

Esperanza Martínez is an Ecuadorian environmentalist, lawyer and biologist committed to the defence of the Earth. She was a co-founder of the renowned environmentalist NGO Accion Ecologica in 1986 and advised the Presidency of the Constituent Assembly of Ecuador in 2007–08. She was also the Coordinator of the International Oilwatch Network between Latin America, Africa and Asia. She is one of the promoters of leaving the oil in the Ecuadorian Yasuní National Park.

Camila Moreno is a Brazilian environmental researcher, activist and author. Her main research interest is the interface between the global environmental governance regime and the greening of capitalism, with a critique of carbon metrics and decarbonisation. She is a founding member of the Grupo Carta de Belém, a Brazilian civil society network against the financialisation of nature.

Zo Randriamaro is an ecofeminist activist-researcher from Madagascar, holds expertise in gender, development sociology and global governance. With a background in sociology and anthropology, she's led international research initiatives, authored works on the gender dimensions of ecological and economic policies, and advised UN agencies. As founder of the Research and Support Center for Development Alternatives, she's devoted to fostering ecofeminist alternatives to development in Africa, in solidarity with other sisters and allies.

Tatiana Roa Avendaño is a Colombian environmentalist, activist and founding member of Censat Agua Viva. She is involved with various organisations and networks advocating for environmental justice, including Oilwatch and the Eco-social and Intercultural Pact of the South. An engineer by training, Tatiana holds a master's degree in Latin American studies and is a PhD candidate at CEDLA–University of Amsterdam. She has written several articles and books on extractivism, water justice and energy sovereignty.

Maristella Svampa is an Argentinian sociologist, researcher and writer. She has received several awards and recognitions, including the Guggenheim Fellowship (2007), the Platinum Konex Award in Sociology (2016), and the National Sociological Essay Award for her book *Latin American Debates. Indianism, Development, Dependence and Populism* (2018). A prolific writer, her research addresses the socio-ecological crisis, social movements, collective action and problematics linked to critical thinking and Latin American social theory.

Ivonne Yanez is a co-founding member of Acción Ecológica, one of the most well-known and respected climate justice organisations in Latin America. She co-founded OilWatch in 1996, an oil activities resistance network. She works on energy, climate change and environmental services from a critical perspective. She has been an active promoter of the Keep the Oil in the Soil campaign for many years, with the Ecuadorian Yasuní National Park as the emblematic case.

Index

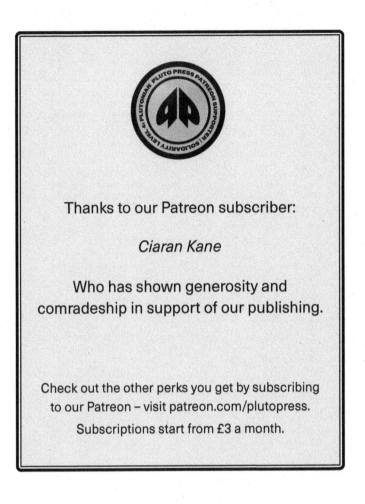

Thanks to our Patreon subscriber:

Ciaran Kane

Who has shown generosity and comradeship in support of our publishing.

Check out the other perks you get by subscribing to our Patreon – visit patreon.com/plutopress.
Subscriptions start from £3 a month.

The Pluto Press Newsletter

Hello friend of Pluto!

Want to stay on top of the best radical books
we publish?

Then sign up to be the first to hear about our
new books, as well as special events,
podcasts and videos.

You'll also get 50% off your first order with us
when you sign up.

Come and join us!

Go to bit.ly/PlutoNewsletter